LAB MANUAL

To accompany

Electricity:

Fundamental Concepts
and Applications

by

Timothy J. Maloney
with
Ben Bartlett and
Stephen Kuchler

Delmar Publishers Inc.®

NOTICE TO THE READER

Publisher does not warrant or guarantee any of the products described herein or perform any independent analysis in connection with any of the product information contained herein. Publisher does not assume, and expressly disclaims, any obligation to obtain and include information other than that provided to it by the manufacturer.

The reader is expressly warned to consider and adopt all safety precautions that might be indicated by the activities described herein and to avoid all potential hazards. By following the instructions contained herein, the reader willingly assumes all risks in connection with such instructions.

The publisher makes no representations or warranties of any kind, including but not limited to, the warranties of fitness for particular purpose or merchantability, nor are any such representations implied with respect to the material set forth herein, and the publisher takes not responsibility with respect to such material. The publisher shall not be liable for any special, consequential, or exemplary damages resulting, in whole or in part, from the readers' use of, or reliance upon, this material.

For information, address Delmar Publishers Inc.
2 Computer Drive West, Box 15-015
Albany, New York 12212

COPYRIGHT © 1992
BY DELMAR PUBLISHERS INC.

All rights reserved. No part of this work covered by the copyright hereon may be reproduced or used in any form or by any means - graphic, electronic, or mechanical, including photocopying, recording, taping, or information storage and retrieval systems - without written permission of the publisher.

Printed in the United States of America
Published simultaneously in Canada
by Nelson Canada
A Division of The Thomson Corporation

10 9 8 7 6 5 4 3 2

Library of Congress Catalog Card Number: 91-30637
ISBN: 0-8273-5040-6

CONTENTS

EXPERIMENT	TITLE	PAGE
1	Electrical Safety Information Sheets	1
2	Constructing a Printed Circuit Board	13
3	Resistor Color Code	17
4	Using the Volt-Ohm-Milliammeter	21
5	Using the Digital Multimeter	25
6	Measuring Resistance with the Volt-Ohm-Milliammeter	27
7	Measuring Resistance using the Digital Multimeter	31
8	Measuring Resistance of Conductors, Insulators, and Semiconductors	35
9	Measuring Components to Determine Defective Components	39
10	Measuring DC Current with Volt-Ohm-Milliammeter	45
11	Measuring DC Current with Digital Multimeter	49
12	Voltage Measurements with the Volt-Ohm-Milliammeter	53
13	Voltage Measurements with a Digital Multimeter	57
14	Batteries — Aiding and Opposing	61
15	Using the Adjustable DC Power Supply	65
16	Ohm's Law	69
17	Simple Series Circuits	73
18	Measuring Resistance in Series Circuits	77
19	Voltage Measurements in Series Circuits	81
20	Measuring Current in Series Circuits	85
21	Connecting Series Circuits Using Ground	89

EXPERIMENT	TITLE	PAGE
22	Voltage Dividers (Unloaded)	95
23	Series Circuits Design Considerations	99
24	Troubleshooting Series Circuits Using Resistance Readings	103
25	Troubleshooting Series Circuits Using Voltage Readings	107
26	Troubleshooting Series Circuits Current Readings	111
27	Series Circuit Troubleshooting Performance Test	115
28	Maximum Transfer of Power	119
29	Analyzing Parallel Circuits	123
30	Troubleshooting Parallel Circuits	125
31	Designing Parallel Circuits	129
32	Analyzing Series–Parallel Combination Circuits	133
33	Loaded Voltage Dividers	137
34	Troubleshooting Series–Parallel Circuits Using Resistance Measurements	141
35	Troubleshooting Series–Parallel Circuits Using Voltage Measurements	143
36	Troubleshooting Series–Parallel Circuits Using Current Measurements	145
37	Troubleshooting Performance Test for Series–Parallel Circuits	147
38	Capacitor Color Code and RC Time Constants	149
39	Inductors and RL Time Constants	153
40	Operation of DC Relays	155
41	AC Voltage	159
42	The Oscilloscope: Familiarization and Setup	163
43	The Oscilloscope: DC Measurements	167

44	The Function Generator (SFG)	171
45	Oscilloscope with Non-Sine Wave Inputs	175
46	Transformers	179
47	Transformer—AC Coupling	183
48	Inductive Reactance	185
49	Inductive Voltage Divider	189
50	Series RL Circuits with Changes in Inductance	191
51	Series RL Circuits with Changes in Frequency	195
52	Parallel RL Circuits with Changes in Inductance	199
53	Parallel RL Circuits with Changes in Frequency	203
54	Capacitive Reactance	207
55	Capacitive Voltage Divider	211
56	Series RC Circuits with Changes in Capacitance	213
57	Series RC Circuits with Changes in Frequency	217
58	Parallel RC Circuits with Changes in Capacitance	221
59	Parallel RC Circuits with Changes in Frequency	225
60	Series LRC Circuits	229
61	Series Resonant Circuits	233
62	Parallel LRC Circuits	237
63	Parallel Resonant Circuits	241
64	Phase Angle Measurement Using the Oscilloscope	245
65	Diode—Resistive Characteristics	249
66	Half-Wave Rectification	251
67	Full-Wave Bridge Rectifier	255
68	Half-Wave Rectifier with Capacitor Filter	259
69	Bipolar Transistor - Resistive Characteristics	263
70	Bipolar Transistor - DC Operation	267

71	Bipolar Transistor Signal Amplifier	271
72	JFET—Resistive Characteristics	275
73	JFET—DC Operation	279
74	JFET—Voltage Amplifier	283

PARTS KIT FOR ELECTRONIC TECHNOLOGY

RESISTORS, 1/4 OR 1/2 WATT

10 ohms	2.2 kilohms	270 kilohms
15 ohms	2.7 kilohms	470 kilohms
22 ohms	3.3 kilohms	560 kilohms
27 ohms	3.9 kilohms	1 megohms
33 ohms	4.7 kilohms	2.2 megohms
39 ohms	5.1 kilohms	3.9 megohms
47 ohms	5.6 kilohms	4.7 megohms
56 ohms	6.2 kilohms	5.1 megohms
68 ohms	6.8 kilohms	
82 ohms	8.2 kilohms	
100 ohms	10 kilohms	
120 ohms	12 kilohms	
150 ohms	15 kilohms	
180 ohms	18 kilohms	
220 ohms	22 kilohms	
270 ohms	33 kilohms	
330 ohms	39 kilohms	
390 ohms	47 kilohms	
470 ohms	56 kilohms	
560 ohms	62 kilohms	
680 ohms	68 kilohms	
820 ohms	75 kilohms	
1 kilohms	100 kilohms	
1.2 kilohms	120 kilohms	
1.5 kilohms	150 kilohms	
1.8 kilohms	220 kilohms	

POTENTIOMETERS

1 kilohm pot
10 kilohm pot
100 kilohm pot

CAPACITORS

390 pF	1 µF
1 nF	2.2 µF
2.7 nF	10 µF
5 nF	22 µF
6.8 nF	100 µF
10 nF	220 µF
15 nF	1000 µF
20 nF	
50 nF	
100 nF	
270 nF	

PARTS KIT FOR ELECTRONIC TECHNOLOGY

TRANSISTORS
2N3904

2N3906

2N6004

2N6005

DIODES
1N4001

1N914/1N4148

LEDs
RED 30 mA

Experiment 1

ELECTRICAL SAFETY

DISCUSSION

Working safely in the lab is the number one goal of the student and the instructor. If a conflict ever arises between doing a project in a safe manner or doing the project more quickly, or more easily, but less safely, ALWAYS decide on the safer approach.

There is NEVER a good enough excuse for you to ignore proper safety procedures for the sake of saving some time, or making the job easier, or for any other reason. Remember that one mistake is all you may get. That mistake could be fatal, or could cause injury or disability. Don't take that chance!

TEXT CORRELATION

Before beginning this experiment, review section 3–7 on factors that determine resistance in *Electricity*.

OBJECTIVES

1. To learn safe practices to use when working in the electronics environment.

2. To learn the correct operation of and location of fire extinguishers in the shop area.

EQUIPMENT

☐ Safety test information sheets concerning electrical saftey fire extinguisher

PROCEDURES	FINDINGS	CONCLUSIONS
1. Read the Electrical Safety Information sheets.		
2. Discuss the safety requirements with the instructor, or attend a safety lecture/demonstration.		
3. Locate and demonstrate the operation of the fire extinguisher in your lab area.		
4. Take the safety test, and pass with 100% grade.		

Experiment 2

CONSTRUCTING A PRINTED CIRCUIT BOARD

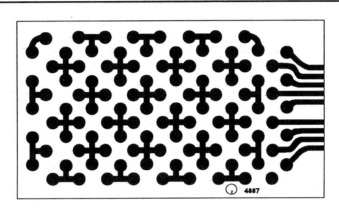

DISCUSSION

The printed circuit board is a relatively new concept in electronics, having only been developed in the last few years. It replaces the older technology of point to point wiring that was prevalent in the days of vacuum tubes. The printed circuit board serves two main functions; the first is as a support upon which the components are mounted. The second function is to electrically connect each component to the other components to form an electronic circuit. Technicians quite often are called upon to build an electronic circuit on a PC board. This can be done with commercially available prototype PC boards, or you can "do it yourself," using the materials listed.

TEXT CORRELATION

Before beginning this experiment, review section 3–8 on conductors and insulators in *Electricity*.

OBJECTIVES

1. To build a prototype PC board, with a standard cloverleaf style track layout.
2. To construct electric circuits using the prototype PC board.

EQUIPMENT

- ☐ Safety glasses, preferably with side shields
- ☐ One 4" X 8" copper clad board, with copper on one or both sides
- ☐ Assorted drafting aids, donuts, and resist tape for masking circuitry on PC boards
- ☐ Ferric chloride acid bath for etching PC board
- ☐ One drill press or portable drill
- ☐ Assorted small drill bits for drilling component holes
- ☐ Tongs for handling PC board
- ☐ Tin plating solution (optional)

14 ■ Constructing a Printed Circuit Board

Experiment 2

SAFETY NOTES

 Ferric chloride is an acid solution. It can do considerable damage to clothing, as well as your eyes. Be very careful handling the acid bath. Acid must be disposed of using accepted hazardous substance procedures.

PROCEDURES	FINDINGS	CONCLUSIONS
1. Using assorted donuts, resist tape, etch resist pen, and other drafting aids, lay out the tracks for the printed circuit as shown in Figure 1.	1. Observe process every 3 or 4 minutes to determine whether the copper is completely etched away from the copper clad board.	1. What is the purpose of the resist tape and donuts used before you etched the PC board? _____ _____ _____ _____.
2. Put on your safety glasses. Carefully insert the PC board into the ferric chloride solution, and agitate gently, with a pair of tongs. Be careful not to splash any of the liquid onto your clothes or in your eyes. Do NOT use your bare hands.	2. The amount of time will vary from 15 or 20 minutes to much longer, depending on the strength of the solution, the temperature of the solution, and the amount of contamination of the solution.	—
3. When all copper is etched away, except that which is under the resist substances, remove the PC board from the acid solution using tongs, and rinse under cold water for 3 minutes. Using abrasive cloth, steel wool, and scouring powder, clean the etch resist from the tracks.	—	—
4. Using the appropriate drill bits, drill the holes in the PC board for the components to be inserted through. The hole size will depend upon the inside diameter of the donuts and/or the size of the leads on the components to be installed in the PC board.	—	4. How many components can be connected to the same track on the PC board? _____ _____ _____ _____.

Experiment 2

Constructing a Printed Circuit Board ▪ 15

PROCEDURES	FINDINGS	CONCLUSIONS
5. If the board is to be tin plated to prevent the tracks from corroding and to improve the solderability of the tracks, immerse the PC board into the tin plating solution for the necessary time to complete this process. (Refer to the instructions on the tin plating kit.)	5. Handle the tin plating solution carefully to avoid splashing or spilling the liquid.	—
6. The PC board is now ready for use. (Note: The pattern shown in Figure 1 is a standard prototype board that may be used to assemble the components and circuitry throughout the remainder of this manual.)	—	—

Experiment 3

RESISTOR COLOR CODE

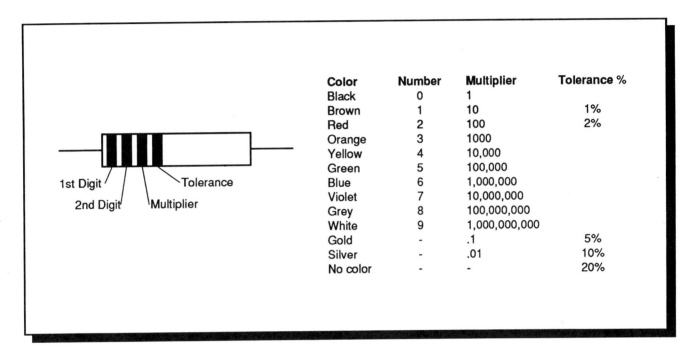

Color	Number	Multiplier	Tolerance %
Black	0	1	
Brown	1	10	1%
Red	2	100	2%
Orange	3	1000	
Yellow	4	10,000	
Green	5	100,000	
Blue	6	1,000,000	
Violet	7	10,000,000	
Grey	8	100,000,000	
White	9	1,000,000,000	
Gold	-	.1	5%
Silver	-	.01	10%
No color	-	-	20%

DISCUSSION

Many components, resistors as well as capacitors and inductors (to name a few), use a color code to designate the nominal value of the component, as well as to indicate the tolerance of the component. That way, the technician can identify the component, determine what the value of the component is, and test the component to determine if it is good or bad.

TEXT CORRELATION

Before beginning this experiment, review section 3–4 on color code in *Electricity*.

OBJECTIVES

1. To use the resistor color code to identify resistors.
2. To measure the value of resistance to determine whether the resistor is good or bad.

EQUIPMENT

☐ Standard parts kit

☐ Digital multimeter, with 10 megohm input resistance on voltage scales OR

☐ Volt-ohm-milliammeter, 20 kilohm/volt DC

18 ■ Resistor Color Code

Experiment 3

PROCEDURES	FINDINGS	CONCLUSIONS

1. Select a resistor from the standard parts kit, with red, red, brown, silver color code, or, if not found, red, red, brown, gold color code. Using the color code chart for resistors (Figure 1), identify the resistor by size (ohmic value) and tolerance rating.

2. Calculate the maximum and minimum value of resistance allowable for the resistor.

3. Using the digital multimeter or volt-ohm-milliammeter, as directed by the instructor, measure the resistor to determine whether it is in tolerance.

   Note: It will be necessary to select the appropriate scale to measure the resistor on. Otherwise, the resistor may measure as a short circuit or open circuit when there is nothing wrong with the resistor.

4. Select the resistor with the yellow, violet, black, gold or yellow, violet, black, silver color code. Once again, use the color code chart (Figure 1) to determine the nominal value.

---

1. Identify the resistor by size (ohmic value) and tolerance rating. Record this value in the blanks below:

   Ohms _____

   Tolerance _____

2. Record the maximum and minimum resistance value allowable for this resistor in the blanks below.

   R_{max} _____

   R_{min} _____

3. Record the measured value of resistance in the blank below:

   R _____

4. Identify the resistor by size (ohmic value) and tolerance rating. Record this value in the blanks below:

   Ohms _____

   Tolerance _____

—

—

3. Is the resistor's measured value within the range of resistances you calculated for this resistor? _____

_____
_____
_____.

4. If a resistor is color coded yellow, violet, black, gold, what is its resistance? _____

_____
_____
_____.

What is the range of resistance values it can read and still be in tolerance? _____

_____
_____
_____.

Experiment 3   Resistor Color Code ■ 19

PROCEDURES	FINDINGS	CONCLUSIONS
5. Calculate the maximum and minimum value of resistance acceptable for the resistor.	5. Record the maximum and minumum resistance for the resistor in the blanks below: R_{max} _____ R_{min} _____	5. How would a resistor of 2.7 ohms, 10% tolerance be coded? _____ _____ _____. What would change for a resistor of .27 ohms, 10% tolerance? _____ _____ _____.
6. Using the digital multimeter or volt-ohm-milliammeter, as directed by the instructor, measure the resistor to determine whether it is in tolerance. Note: It will be necessary to select the appropriate scale to measure the resistor on. Otherwise, the resistor may measure as a short circuit or open circuit when there is nothing wrong with the resistor.	6. Record the measured value of resistance in the blank below: R _____	6. Is the resistor within the range of resistance readings you determined to be normal? _____ _____ _____.
7. Select the resistor with the orange, orange, yellow, gold or orange, orange, yellow, silver color code. Once again, use the color code chart (Figure 1) to determine the nominal value.	7. Identify the resistor by size (ohmic value) and tolerance rating. Record the value in the blanks below: Ohms _____ Tolerance _____	7. How much current would a resistor that is color coded orange, orange, brown, gold have through it if it were connected across a voltage source of 10 volts? _____ _____ _____.
8. Calculate the maximum and minimum value of resistance acceptable for the resistor. —	8. Record the maximum and mimumum resistance for the resistor in the blanks below: R_{max} _____ R_{min} _____ 9. Record the measured value of resistance in the blank below: R _____	— —

Experiment 4

USING THE VOLT-OHM-MILLIAMMETER

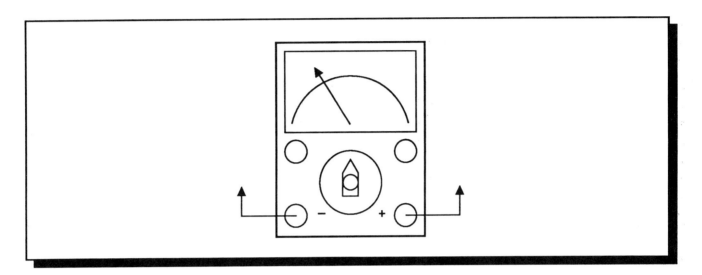

DISCUSSION

Even though the volt-ohm-milliammeter (VOM) has been around for many years and does not represent the latest technology, it is still the meter of choice for many technicians for many jobs. It remains the best meter for indicating peak values, as in performing an alignment of a receiver. It is considered by many technicians to be the best choice for evaluating transistor operation. Also, many of the more experienced technicians have made additional test fixtures to make the VOM into a wattmeter, or pulse counter, etc.

So, it is essential to learn how to use the VOM, because it will probably still be used by the industry for many years to come.

TEXT CORRELATION

Before beginning this experiment, review section 5–4 on multimeters in *Electricity*.

OBJECTIVES

1. To learn the basic operation of the VOM, and examine the operating controls.

2. To evaluate the functions of the VOM.

EQUIPMENT

☐ Volt-ohm-milliammeter, 20 kilohm/volt DC

☐ One black test lead

☐ One red test lead

22 ■ Using the Volt-Ohm-Milliammeter   Experiment 4

PROCEDURES	**FINDINGS**	**CONCLUSIONS**

1. Examine front panel of the VOM asigned to you. Locate the function switch. This will be a multifunction switch with several positions for measuring VOLTS, OHMS, & AMPS. In the blanks indicated, record the selections available.

1. Record the functions available on your multimeter.

VOLTS	OHMS	AMPS
___	___	___
___	___	___
___	___	___
___	___	___
___	___	___
___	___	___
___	___	___

1. In your own words, explain why there are so many different scales on the VOM for measuring current, rather than just one scale capable of measuring the largest amount of current the manufacturer designed the meter to measure.

2. Jacks are located on the front panel for insertion of the test leads. Insert the black test lead into the jack marked "Common -", or maybe just "-".

—

—

3. Insert the red test lead into the jack marked "Volts/Ohms" or simply "+". These leads will be used for connecting to circuitry and/or components.

—

—

4. Examine the meter face. There should be several scales with numbers on them. In the blanks indicated, record the scales available. (Note: There usually is one scale for resistance, located at the top of the meter face, and several scales for AC and DC voltages located below the resistance scale.)

4. Record the scales available for the meter you are using.

DC VOLTS	AC VOLTS	CURRENT	OHMS
___	___	___	___
___	___	___	___
___	___	___	___
___	___	___	___
___	___	___	___

—

5. There may also be a front panel switch for selecting polarity and type of voltage measurement. This will be labeled "- DC", "+ DC", and "AC". If your VOM has this type of switch, place it in the "+ DC" position.

—

—

6. On the meter face, there will be a screwdriver adjustment for zeroing the meter movement. Locate this adjustment, and adjust the meter movement to zero on left side of meter.

—

6. What might be the effect on the readings obtained with the VOM if step 6 is ignored?

_____

_____

Experiment 4

Using the Volt-Ohm-Milliammeter ■ 23

PROCEDURES	FINDINGS	CONCLUSIONS

_____
_____
_____

Does your VOM have a different kind of scale for resistance readings and voltage readings? _____
If so, explain what the difference is between them.

_____
_____
_____
_____

7. Place meter function switch on lowest OHMS position (usually R x 1). Touch both test leads together, and observe movement of needle. It should move to right side of scale. Adjust for zero on right side using the "Zero Ohms" adjust on meter.   —   —

8. When finished, move the function switch to the highest voltage setting. This will preserve the batteries in the meter as well as protect against accidental damage to the meter if the next technician forgets to change the function switch setting.   —   —

Experiment 5

USING THE DIGITAL MULTIMETER

DISCUSSION

The digital multimeter (DMM) is a fairly new instrument, having only come into widespread use in the last ten or fifteen years. Each year, as the new models are brought out, more and more features are added, and more tests can be run with them. The VOM is gradually being replaced for many of the tests it used to excel at, although there are still many uses for it. The DMM is much more accurate, and has much less effect on the circuit (loading effect and insertion effect) than the VOM.

TEXT CORRELATION

Before beginning this experiment, review section 5-4 on multimeters in *Electricity*.

OBJECTIVES

1. To examine the front panel controls of the DMM, and determine what functions can be performed.

EQUIPMENT

- ☐ Standard parts kit
- ☐ Digital multimeter, with 10 megohm input resistance on voltage scales
- ☐ One black test lead for the meter
- ☐ One red test lead for the meter

26 ■ Using the Digital Multimeter   Experiment 5

PROCEDURES	FINDINGS	CONCLUSIONS

1. Examine the front panel of the digital multimeter assigned to you. Locate the function switch or switches. This is the switch that selects whether VOLTS, OHMS, or AMPS are to be read on the meter face. (Note: Many DMMs have several other functions, such as transistor or diode testing, capacitor testing, frequency counting, etc. We will concern ourselves only with VOLTS, OHMS, and AMPS in this section.)

   —

   1. Why is it necessary to provide the user of the DMM with so many scales for measuring voltage, rather than one large scale capable of measuring the largest voltage the manufacturer designed the DMM to read?

   _____
   _____
   _____
   _____

2. Record the selections available in blanks indicated.

   2. Record the selections available in the blanks below:

VOLTS	OHMS	AMPS
____	____	____
____	____	____
____	____	____
____	____	____
____	____	____
____	____	____
____	____	____

   2. What is the biggest advantage, in your opinion, of the DMM compared to the VOM?

   _____
   _____
   _____
   _____

3. Insert the black test lead into the jack labeled "Common -" on the front panel.

   3. Some meters simply call this jack "-" or "Common".

   3. What is the biggest advantage, in your opinion, of the VOM compared to the DMM?

   _____
   _____
   _____
   _____

4. Insert the red test lead into the "Volts/Ohms" jack on the front panel.

   4. Some meters simply call this test jack "+".

   —

5. Rotate the Function Switch to lowest resistance setting.

   —

   —

6. Connect test leads together and observe reading on the meter face. Record the reading in the blank indicated.

   6. Record the reading obtained in the blank below:

   _____
   _____
   _____

   —

Experiment 6

MEASURING RESISTANCE WITH THE VOLT-OHM-MILLIAMMETER

DISCUSSION

The VOM is often used to measure the resistance of a circuit, to determine whether the circuit is functioning normally or whether there is a short circuit or open circuit. Most VOMs have a resistance scale on the meter face at the top. Usually there will be only one scale, plus a function switch that tells the technician the number to multiply the indications on the meter face by. So, if the needle is pointing to 50 on the ohms scale, and the function switch is on R X 1, the resistance is 50 ohms (50 X 1 = 50). If the function switch were on R X 100, the resistance of the device under test is 5000 ohms (50 X 100 = 5000). This takes a little getting used to, but, with practice, it can be mastered.

TEXT CORRELATION

Before beginning this experiment, review section 5–4 on multimeters in *Electricity*.

OBJECTIVES

1. To set up and adjust the VOM to read resistance in ohms.

2. To use the VOM to measure the resistance of various resistors.

EQUIPMENT

☐ Standard parts kit

☐ Volt-ohm-milliammeter, 20 kilohm/volt DC

PROCEDURES	FINDINGS	CONCLUSIONS
1. Insert the banana plug end of the black test lead into the meter's front panel jack labeled "Common".	1. On some meters this jack is simply labeled "–".	1. What scale would you place the meter on to measure for a short circuit? _____ Why?_____.
2. Insert the banana plug end of the red test lead into the front panel jack labeled "+".	2. On some meters this jack is labeled "Volts/Ohms".	2. What scale would you place the meter on to measure for an open circuit? _____ Why?_____.
3. Observe the position of the meter pointer. The pointer should be exactly aligned with the "Infinity" symbol on the left side of the meter face. (Note: This will correspond to the 0 reading on any of the voltage scales.)	3. The "Infinity" symbol looks like a number eight laying on its side.	3. If the 10 ohm resistor were open, what would the reading have been in step 9?_____. Why?_____. If the 47 ohm resistor were open, what would the reading have been in step 9?_____. Why?_____.
4. If the pointer is not aligned properly, adjust the mechanical zero with a small screwdriver.	4. Be careful not to force the adjustment beyond its limits, or to be rough in performing the adjustment. The meter movement could be damaged.	—
5. Place the function switch on the lowest resistance scale. On most VOMs this will be R X 1. This simply means R times one, or, in other words, the numbers on the resistance scale are to be read directly.	5. On most meters, the ohms scale will be the top scale. Zero is to the extreme right on this scale, and infinity is to the extreme left.	—

Experiment 6
Measuring Resistance with the Volt-Ohm-Milliameter

PROCEDURES	FINDINGS	CONCLUSIONS
6. Connect the red lead to the black lead.	—	—
7. Observe the reading on the VOM meter face. It should be near zero, on the RIGHT side of the meter face. If the reading is not zero, adjust for a reading of zero ohms using the "Ohms Adjust" or "Zero Ohms Adjust" on the front panel of the meter.	7. Record the reading obtained in the blank below: _____ _____ _____ _____.	—
8. Place the meter leads across the leads of the 47 ohm resistor in your standard parts kit.	8. Record the reading obtained in the blank below: _____ _____ _____ _____.	—
9. Repeat step 8 for the 10 ohm resistor.	9. Record the reading obtained in the blank below: _____ _____ _____ _____.	—
10. Repeat step 8 for the 150 ohm resistor.	10. Record the reading obtained in the blank below: _____ _____ _____ _____.	—
11. Place the meter on a higher scale that will allow you to read a 1 kilohm resistor.	—	—
12. Repeat steps 6 and 7 to zero the meter on the new scale.	—	—
13. Repeat steps 8 and 9 for the 1 kilohm, 4.7 kilohm, and 10 kilohm resistors.	13. Record the reading obtained in the blank below: _____1 kilohm _____4.7 kilohm _____10 kilohm	—

PROCEDURES	FINDINGS	CONCLUSIONS
14. Repeat this procedure for the 22 kilohm, 47 kilohm, 150 kilohm, 2.2 megohm, and 4.7 megohm resistors, changing scales as appropriate to obtain the best readings. (Note: The best accuracy on the VOM resistance scales will be in the lowest half of the meter face—the right side of the meter face.	14. Record the reading obtained in the blank below: _____ 22 kilohm _____ 47 kilohm _____ 150 kilohm _____ 2.2 megohm _____ 4.7 megohm	—

Experiment 7

MEASURING RESISTANCE USING DIGITAL MULTIMETER

DISCUSSION

The most common test performed by the technician in analyzing electronic circuitry is the resistance test. With the ohmmeter, it can easily be determined whether there is a short circuit or an open circuit. The resistance test is also the test that is normally used to "verify" a failure. In other words, if the technician, through a series of tests performed on the equipment, decides that there is a short or open circuit he/she will then deenergize the circuit, and perform a resistance test of the component to verify if it is shorted or open.

TEXT CORRELATION

Before beginning this experiment, review section 5–4 on multimeters in *Electricity*.

OBJECTIVES

1. To perform resistance tests with the digital multimeter.
2. To perform proper steps in setting up the digital multimeter.

EQUIPMENT

☐ Standard parts kit

☐ Digital multimeter, with 10 megohm input resistance on voltage scales

32 ■ Measuring Resistance Using Digital Multimeter	Experiment 7

PROCEDURES	FINDINGS	CONCLUSIONS

1. Insert the banana plug end of the black test lead into the meter's front panel jack, labeled "Common".

   —

   —

2. Insert the banana plug end of the red test lead into the front panel jack labeled "+".

   —

   —

3. Place the function switch on the lowest resistance scale.

   —

   3. What scale would you place the meter on to measure for a short circuit?_____
   Why?_____
   _____
   _____

4. Connect the red lead to the black lead.

   —

   What scale would you place the meter on to measure for an open circuit?_____
   Why?_____
   _____
   _____

5. Observe the reading on the DMM meter face. It should be near zero. If it is not, check your setup. Some DMMs include a "Zero Adjust" on the front panel. Most do not, choosing instead to set the meter for zero internally. If no problem is found (dead battery, blown fuse, defective meter lead, etc.), notify the instructor.

   5. Most DMM will read a few tenths of an ohm of resistance. This simply represents the resistance of the test leads, and should be considered "zero" for your meter.

   Record the reading obtained in the blank below:

   _____
   _____
   _____
   _____

   —

6. Place the meter leads across the leads of the 47 ohm resistor in your standard parts kit. Refer to the resistor color code to aid in identifying each resistor.

   6. Record the reading obtained in the blank below:

   _____
   _____
   _____
   _____

   —

Experiment 7 Measuring Resistance Using Digital Multimeter ■ 33

PROCEDURES	FINDINGS	CONCLUSIONS
7. Repeat steps 6 and 7 for the 10 ohm resistor.	7. Record the reading obtained in the blank below: _____ _____ _____ _____ .	—
8. Repeat steps 6 and 7 for the 150 ohm resistor.	8. Record the reading obtained in the blank below: _____ _____ _____ _____ .	8. If the 10 ohm resistor were open, what would the reading have been in step 9?_____ Why?_____ _____. If the 47 ohm resistor were open, what would the reading have been?_____ Why?_____ _____.
9. Place the meter on a higher scale that will allow you to read a 1 kilohm resistor.	—	—
10. Repeat step 5 to zero the meter on the new scale.	—	—
11. Repeat steps 6 and 7 for the 1 kilohm, 4.7 kilohm, and 10 kilohm resistors.	11. Record the reading obtained in the blank below, for the 1 kilohm resistor: _____ _____ . Record the reading obtained in the blank below, for the 4.7 kilohm resistor: _____ _____ . Record the reading obtained in the blank below, for the 10 kilohm resistor: _____ _____ _____ .	—

PROCEDURES	FINDINGS	CONCLUSIONS
12. Repeat this procedure for the 22 kilohm, 47 kilohm, 150 kilohm, 2.2 megohm and 4.7 megohm resistors, changing scales as appropriate to obtain the best readings.	12. Record the reading obtained in the blank below, for the 22 kilohm resistor: _____ _____ _____ _____ Record the reading obtained in the blank below, for the 47 megohm resistor: _____ _____ _____ _____ Record the reading obtained in the blank below, for the 150 megohm resistor: _____ _____ _____ _____ Record the reading obtained in the blank below, for the 2.2 megohm resistor: _____ _____ _____ _____ Record the reading obtained in the blank below, for the 4.7 megohm resistor: _____ _____ _____ _____	—

Experiment 8

MEASURING RESISTANCE OF CONDUCTORS, INSULATORS, AND SEMICONDUCTORS

DISCUSSION

The ohmmeter is used to determine the condition of many components, to see whether they are shorted, or open, or normal. Therefore, it is necessary to know what the normal readings are for various materials in order that the technician will have something to refer to. In this project, we will measure the resistance of some common household items, such as popsicle sticks and rubber bands, and compare the readings to electronic components, such as capacitors and diodes.

TEXT CORRELATION

Before beginning this experiment, review section 5–4 on multimeters in *Electricity*.

OBJECTIVES

1. To measure the resistance of conductors, insulators, and semiconductors.

2. To determine what resistance readings are associated with conductors, insulators, and semiconductors.

EQUIPMENT

- ☐ Digital multimeter, with 10 megohm input resistance on voltage scales OR

- ☐ Volt-ohm-milliammeter, 20 kilohm/volt DC

- ☐ Assorted items to be measured, including: popsicle stick, rubber band, short length of hookup wire (#20 or #22 gauge), short length of nichrome wire, silicon rectifier 1N4001 electrolytic capacitor 100 µF

36 ■ Measuring Resistance of Conductors, Insulators and Semiconductors Experiment 8

PROCEDURES	**FINDINGS**	**CONCLUSIONS**
1. Set up the volt-ohm-milliameter or the digital multimeter (as directed by your instructor) to measure resistance.	—	—
2. Select first item, the popsicle stick, from the list of items to be measured.	—	—
3. Measure the resistance value of the popsicle stick, on the lowest scale available on the meter. If the reading is too high for that scale, change the setting of the function switch to a higher scale, until the reading will fit on the scale. If it is too high for even the highest range available, reverse the connection of the leads to the popsicle stick, and observe the reading. If the reading is still too high, enter "infinity" in the blank, and list the item as an insulator.	3. Record the reading obtained in the blank below: _____ _____ _____ _____. The measured value indicates the component is a/an _____ _____ insulator/conductor/semiconductor/resistor.	—
4. Measure the resistance value of the rubber band, in the same manner as the popsicle stick.	4. Record the reading obtained in the blank below: _____ _____ _____ _____. The measured value indicates the component is a/an _____ insulator/conductor/semiconductor/resistor.	—
5. Measure the resistance of the hookup wire. You should discover that the resistance reading is low enough so that you will not need to change the meter from the lowest resistance scale. If the reading is low both ways, the item is a conductor, and you would enter "conductor" in the blank.	5. Record the reading obtained in the blank below: _____ _____ _____ _____ The measured value indicates the component is a/an _____ _____ insulator/conductor/semiconductor/resistor.	5. In general, what range of resistance readings would be appropriate for conductors? _____ _____.

Experiment 8　　　　　　　　　　Measuring Resistance of Conductors, Insulators and Semiconductors ■ 37

PROCEDURES	FINDINGS	CONCLUSIONS

6. Measure the resistance of the nichrome wire. You will discover that the resistance is higher than that of the hookup wire, but you may not need to change scales on the meter to measure it. If the reading is constant both ways, but is considerably higher than zero ohms, the item is a resistor and you would enter "resistor" in the blank.

6. Record the reading obtained in the blank below:

_____
_____
_____
_____.

Each time you attempt to read the value of a new component, start on the lowest resistance scale. Otherwise, if you are on a high scale, the component may read as a conductor, when actually it has considerable resistance. For example, if you test the nichrome wire on the highest scale, it will read 0 ohms; but when you change the meter to a more appropriate scale, it can be seen that the nichrome wire actually has much more resistance than a conductor would normally have.

The measured value indicates the component is a/an _____ _____ insulator/conductor/semiconductor/resistor.

—

7. Measure the resistance of the silicon rectifier. If, when you reverse the meter leads, the reading changes to some lower value, or starts at a low value then changes to a higher value when the leads are reversed, the item is a semiconductor, and you would enter "semiconductor" in the appropriate column in Table 1.

7. Record the reading obtained in the blank below:

_____
_____
_____
_____.

The measured value indicates the component is a/an _____ _____ insulator/conductor/semiconductor/resistor.

7. What did you discover about the resistance of the semiconductor diode?

_____
_____.

38 ■ Measuring Resistance of Conductors, Insulators and Semiconductors
Experiment 8

| **PROCEDURES** | **FINDINGS** | **CONCLUSIONS** |

8. Measure the resistance of the capacitor. The capacitor may start out as a conductor, but the reading will gradually change, going higher and higher until it reads as an insulator.

8. Record the reading obtained in the blank below:

_____

_____

_____

_____

The measured value indicates the component is a/an _____ _____ insulator/conductor/semiconductor/resistor

8. You probably got some strange results when testing the electrolytic capacitor. Can you explain what causes these strange readings?

_____

_____

Describe, in your own words, what you have discovered about the resistance of conductors, insulators, semiconductors, and resistors, from performing this experiment.

_____

_____

_____

_____

_____

_____

_____

Experiment 9

MEASURING COMPONENTS TO DETERMINE DEFECTIVE COMPONENTS

DISCUSSION

The final verification test usually performed on a component to be certain of its condition is the resistance test. If the component is shorted, that will show up as a low resistance when the technician performs a resistance test. If the component is open, except in a few cases, that will also show up, as a very high resistance reading. Therefore, the resistance test is one of the most important procedures the technician can learn.

TEXT CORRELATION

Before beginning this experiment, review section 5–4 on multimeters in *Electricity*.

OBJECTIVES

1. To perform resistance tests on good components to verify the condition of the component.

2. To perform resistance tests on defective components to determine the type of defect.

EQUIPMENT

- ☐ Standard parts kit
- ☐ Digital multimeter, with 10 megohm input resistance on voltage scales OR
- ☐ Volt-ohm-milliammeter, 20 kilohm/volt DC
- ☐ One good resistor, capacitor, and diode, chosen by the instructor, to ensure adequate practice reading both good and bad components
- ☐ One resistor, one capacitor, one diode, with defects

Measuring Components to Determine Defective Components

Experiment 9

PROCEDURES	FINDINGS	CONCLUSIONS
1. Set up the digital multimeter or VOM, as directed by the instructor, to measure resistance.	—	1. In your own words, describe the process of determining the condition of a suspected component. _____ _____ _____ _____
2. Select the first item to be tested, which is a diode (one of two diodes). Determine the range of expected good readings, in ohms. Place the meter on an appropriate range to measure this value.	2. The diode should read low resistance in one direction, and high resistance in opposite direction. So when you read the resistance one way, and then reverse your meter leads, the resistance should go from low to high, or from high to low. The ratio should be at least 100:1 of resistance in one direction and resistance in opposite direction.	—
3. Measure the component.	3. Record the reading obtained in the blank below: R low _____ R _____ Condition _____ Note: In blank labeled "condition," the component is either good, shorted, or open.	—
4. If a reading of 0 ohms is obtained, move the range selector switch (function switch) to the next lowest range.	—	—
5. If the reading is still 0 ohms, continue to change the range selector until the range selector is on the lowest resistance range or until a reading of something other than zero is obtained.	—	5. Why is it necessary to change the range switch if your initial reading indicates a short circuit? _____ _____ _____ _____

Experiment 9

Measuring Components to Determine Defective Components ■ 41

PROCEDURES	FINDINGS	CONCLUSIONS

Why is the range switch changed in a downward direction?

_____
_____
_____
_____

6. If the reading is too high for the scale selected, move the function selector to the next highest scale. If it is still too high to be read on the next scale, continue moving the range selector until a reading is obtained or until the highest resistance scale is reached.

—

6. Why is it necessary to change the range switch if your initial reading indicates an open circuit?

_____
_____
_____
_____

Why is the range switch changed in an upward direction?

_____
_____
_____
_____

7. Repeat this procedure for the second diode, recording the results as before.

7. Record the reading obtained in the blank below:

R low _____
R high _____
Condition _____

Note: In blank labeled "condition," the component is either good, shorted, or open.

—

8. Select one of the two capacitors, and perform the procedure as before, recording the results.

8. The capacitor should have a large resistance in both directions (in other words, no matter which way the ohmmeter leads are connected). For this reason, it may not be possible to tell if the capacitor is open or not, especially if the capacitor is a small value capacitor. However, if the capacitor is a

—

PROCEDURES	FINDINGS	CONCLUSIONS
	large value capacitor, 1 µF or larger, it should be possible to determine whether the capacitor is able to charge and discharge through the resistance of the ohmmeter. This will be indicated by a progressively higher and higher resistance reading the longer the meter is connected. If the leads are reversed, the capacitor will discharge through the meter, and charge back up in the opposite direction. However, in the final analysis, the only certain test for an open capacitor, especially a small one, is to replace it and see if the symptoms disappear. It will be possible, however, to determine if a capacitor is shorted. This is indicated by a low resistance reading in both directions. Record the reading obtained in the blank below: _____ Condition_____ Note: In blank labeled "condition," the component is either good, shorted, or open.	
9. Perform the resistance test on the remaining capacitor.	9. Record the reading obtained in the blank below: _____ Condition_____ Note: In blank labeled "condition," the component is either good, shorted, or open.	—
10. Perform the resistance test on the first of the two resistors, and record the results.	10. To determine the normal reading of the resistor, simply determine the maximum and minimum resistance based on the resistor color code. If the color code indicates 5% tolerance, multiply the resistor's	—

PROCEDURES	FINDINGS	CONCLUSIONS
	nominal size by .95 to obtain the lowest acceptable reading, and by 1.05 to determine the highest acceptable reading. If the tolerance is 10%, multiply by .9 and by 1.1. Record the reading obtained in the blank below: _____ Condition_____ Note: In blank labeled "condition," the component is either good, out of tolerance, shorted, or open.	
11. Select the second resistor, and perform the same tests on it, recording the results.	11. Record the reading obtained in the blank below: _____ Note: In blank labeled "condition," the component is either good, out of tolerance, shorted, or open.	—

Experiment 10

MEASURING DC CURRENT WITH VOLT-OHM-MILLIAMETER

SCHEMATIC:

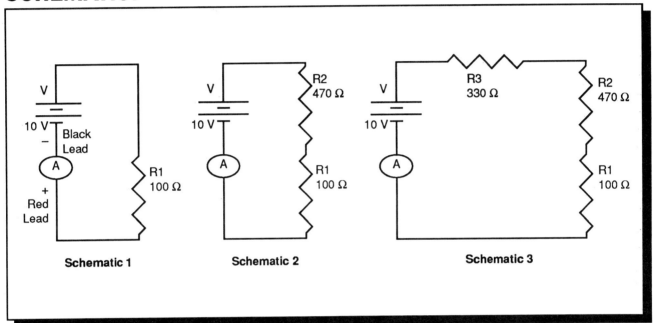

DISCUSSION

The volt-ohm-milliammeter has been in use for many years and is still the meter of choice for many technicians, simply because it remains the best meter to measure certain values. Also, many technicians who have a great deal of experience have developed special ways of extending the usefulness of the VOM, so it can be used to measure many things for which it was not designed. Therefore, it is necessary for the beginning technician to learn to properly operate the VOM, for measuring current, voltage, and resistance.

OBJECTIVES

1. To learn to operate the VOM as a milliammeter.
2. To measure current in a simple electric circuit.

TEXT CORRELATION

Before beginning this experiment, review section 5–1 on measuring current in *Electricity*.

EQUIPMENT

☐ Standard parts kit

☐ Volt-ohm-milliammeter, 20 kilohm/volt DC

☐ Adjustable power supply, capable of 0 to 40 VDC output

46 ■ Measuring DC Current with Volt-Ohm-Milliameter   Experiment 10

SAFETY NOTES

 It is a good idea to set the ammeter to the highest scale, until it can be verified that the current is small enough to be safely read on a lower scale; then reduce range to obtain reading.

PROCEDURES	FINDINGS	CONCLUSIONS

1. Insert the banana plug end of the black test lead into the meter's front panel jack, labeled "Common".

   —

   —

2. Insert the banana plug end of the red test lead into the front panel jack labeled "+".

   —

   —

3. Observe the position of the meter pointer. The pointer should be exactly aligned with the 0 on the left side of the meter face.

   3. If the meter has a mirrored meter face, align the needle (pointer) with its image in the mirror for best accuracy.

   —

4. If the pointer is not aligned properly, adjust the mechanical zero with a small screwdriver.

   4. Be careful not to damage the meter face cover or the small plastic screw. Do not force the adjustment beyond its limits.

   —

5. Calculate the value of current that will flow in the circuit depicted in Schematic 1 and record it. Place the function switch on a high enough scale to measure calculated current.

   5. Record the calculated value in the blank below:

   I_t _____

   _____.

   5. Explain, in your own words, why it is a good practice to begin your measurements of current by setting the meter on the highest scale.

   _____
   _____
   _____
   _____.

6. Connect the circuit shown in Schematic 1. Connect the red lead to the bottom end of the 100 ohm resistor in Schematic 1, and the black lead to "-" end of the battery.

   —

7. Observe the reading on the VOM meter face. Record the reading.

   7. Record the reading obtained in the blank below:

   I_t _____

   _____.

   7. The VOM has only a small amount of resistance when used as an ammeter. Because of this, what precautions must you take to prevent damage to the meter or to the circuit?

   _____
   _____
   _____
   _____.

Experiment 10   Measuring DC Current with Volt-Ohm-Milliameter ■ 47

PROCEDURES	FINDINGS	CONCLUSIONS

8. Repeat steps 5, 6 and 7 for Schematic 2.

8. Record the calculated value in the blank below:

I_t ———————

————————————.

Record the reading obtained in the blank below:

I_t ———————

————————————.

—

9. Repeat procedure for Schematic 3.

9. Record the calculated value in the blank below:

I_t ———————

————————————.

Record the reading obtained in the blank below:

I_t ———————

————————————.

9. Since it is necessary to break the circuit at the point we want to measure current, can you list some shortcuts to taking current measurements in a circuit consisting of a battery, a switch, a fuse in a fuse holder, and a PC board with several series resistors mounted on it?

————————————
————————————
————————————
————————————

Experiment 11

MEASURING DC CURRENT WITH DIGITAL MULTIMETER

SCHEMATIC:

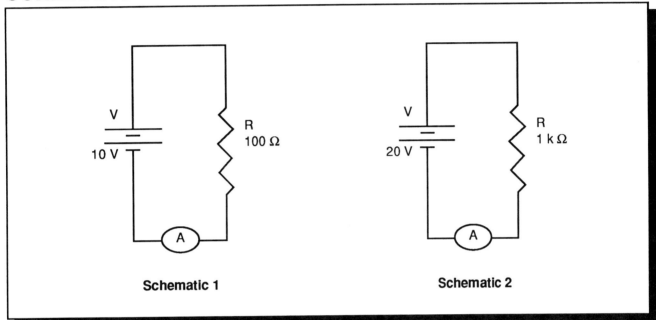

Schematic 1   Schematic 2

DISCUSSION

The digital multimeter has largely replaced the analog multimeter in many electronics applications. Because of its increased accuracy, the digital multimeter can be used whenever it is desirable to obtain a very accurate reading. The digital multimeter is also easier to read, since it displays the information directly, on a liquid crystal display (LCD) readout. Because of its increasing popularity, the technician must learn to measure quantities with it.

TEXT CORRELATION

Before beginning this experiment, review section 5–1 on measuring current in *Electricity*.

OBJECTIVES

1. To measure current in a series circuit using the digital multimeter.

EQUIPMENT

☐ Standard tool box

☐ Standard parts kit

☐ Digital multimeter, with 10 megohm input resistance on voltage scales

☐ Adjustable power supply, capable of 0 to 40 VDC output

50 ■ **Measuring DC Current with Digital Multimeter** Experiment 11

SAFETY NOTES

 It is a good idea to set the ammeter to the highest scale, until it can be verified that the current is small enough to be safely read on the lower scale. You then can place the meter on a lower scale.

PROCEDURES	**FINDINGS**	**CONCLUSIONS**

1. Insert the banana plug end of the black test lead into the meter's front panel jack, labeled "Common".

   —

   —

2. Insert the banana plug end of the red test lead into the front panel jack labeled "AMPS" or "mA".

   —

   —

3. Calculate the value of current that will flow in the circuit depicted in Schematic 1. Place the function switch on a high enough scale to measure calculated current.

   3. Record the calculated value in the blank below:

   I_t _____

   _____

   —

   3. The VOM places a very small amount of resistance in the circuit when you are measuring current. Because of this, what precautions must you take to prevent damage to the meter or to the circuit?

   _____
   _____
   _____
   _____.

4. Connect the circuit shown in Schematic 1. Connect the red lead to the bottom end of the 100 ohm resistor in Schematic 1, and the black lead to "-" end of the battery.

   —

   —

5. Observe the reading on the DMM meter face.

   5. Record the reading obtained in the blank below:

   I_t _____

   _____.

   5. It was explained in the procedure that it is good practice to start on the highest current scale available, then drop down to the lower scale after it is determined that the reading will "fit" that scale. Explain, in your own words, why this is good practice.

   _____
   _____
   _____
   _____.

Experiment 11 | Measuring DC Current with Digital Multimeter ■ 51

PROCEDURES	FINDINGS	CONCLUSIONS
6. Repeat steps 3, 4, and 5 for Schematic 2.	6. Record the calculated value in the blank below: _____ _____ _____ _____ Record the reading obtained in the blank below: _____ _____ _____ _____	6. Breaking the circuit so you can insert an ammeter into the circuit is time consuming, as well as potentially destructive to the PC board. Considering this, if you have a PC board with resistors, a fuse in a fuse holder, a switch to turn the circuit on and off, and a battery, how would you go about measuring current in the most convenient way, with the least chance of damage to the PC board? _____ _____ _____ _____
—	—	7. State, in your own words, the principle you have discovered in this experiment. Explain the relationship between voltage, resistance, and current. _____ _____ _____ _____

Experiment 12

VOLTAGE MEASUREMENTS WITH THE VOM

DISCUSSION

One of the most common measurements taken by the technician is the measurement of voltage. All active devices, transistors, vacuum tubes, and integrated circuits must have DC voltage applied to them to make them operate. The technician compares the voltage readings on the various electrical elements of the transistor, vacuum tube, or integrated circuit to the normal readings for the device under test, to determine whether the device is functioning correctly.

TEXT CORRELATION

Before beginning this experiment, review section 5–2 on measuring voltage in *Electricity*.

OBJECTIVES

1. To set up the VOM to measure DC voltage.

2. To measure the voltage produced by dry cell batteries.

EQUIPMENT

- ☐ Standard tool box
- ☐ Assorted dry cell batteries, one 1.5 volt D cell, one 1.5 volt C cell, and one 9 volt transistor radio battery.
- ☐ Standard parts kit
- ☐ Volt-ohm-milliammeter, 20 kilohm/volt DC

PROCEDURES

1. Insert the banana plug end of the black test lead into the meter's front panel jack, labeled "Common".

2. Insert the banana plug end of the red test lead into the front panel jack labeled "+".

3. Observe the position of the meter pointer. The pointer should be exactly aligned with the 0 on the left side of the meter face.

4. If the pointer is not aligned properly, adjust the mechanical zero with a small screwdriver.

5. Place the function switch on a high enough scale to measure 1.5 volts.

6. Connect the red lead to "+"end of the D battery, and the black lead to "-" end of the battery.

7. Observe the reading on the VOM meter face. Record the reading.

8. Repeat the procedure for the C cell, and record the results in the blanks indicated.

FINDINGS

1. On some meters this jack may be labeled "-".

2. On some meters this jack may be labeled "Volts/Ohms".

—

4. Be certain to use care with this adjustment, since this is a rather easily damaged part.

5. This varies from meter to meter, and will depend on the make and model of your meter.

—

7. Record the reading obtained in the blank below:

_____
_____
_____
_____.

8. Record the reading obtained in the blank below:

_____
_____
_____
_____.

CONCLUSIONS

—

—

—

—

—

6. What would be the effect if the meter leads were reversed in step 6?

_____
_____
_____
_____.

—

8. Compare the readings obtained with the D battery and the C battery. What was the difference in the voltage readings?

_____
_____
_____
_____

Experiment 12   Voltage Measurements with the VOM ■ 55

PROCEDURES	**FINDINGS**	**CONCLUSIONS**

Why are there several sizes of battery, such as D, C, AA, AAA, if all have the same output voltage?

_____
_____
_____
_____.

9. Place the function switch on a high enough scale to allow a reading of at least 9 volts.

— —

10. Repeat the procedure for the transistor radio battery, and record the results in the blanks indicated.

10. Record the reading obtained in the blank below:

_____
_____
_____
_____.

10. Why is the voltage from the transistor radio battery so much higher than the D cell, since the D cell is so much larger in physical size?

_____
_____
_____
_____.

Experiment 13

VOLTAGE MEASUREMENTS WITH A DIGITAL MULTIMETER

DISCUSSION

The digital multimeter, usually referred to as a DMM, is an essential part of the technician's tool kit. Much more accurate than the volt-ohm-milliammeter, it is also easier to read. Voltage readings with this type of multimeter are used very often in troubleshooting and analysis of electronic equipment.

TEXT CORRELATION

Before beginning this experiment, review section 5–2 on measuring voltage in *Electricity*.

OBJECTIVES

1. To measure voltages produced by some common sizes of dry cell batteries.
2. To set up the digital multimeter to measure the voltages.

EQUIPMENT

☐ Digital multimeter, with 10 megohm input resistance on voltage scales

☐ Assorted dry cell batteries, one 1.5 volt D cell, one 1.5 volt C cell, and one 9 volt transistor radio battery

58 ■ Voltage Measurements with a Digital Multimeter

Experiment 13

SAFETY NOTES

 Although the voltages you will be measuring are very low voltages, you are building habits to be used later with high voltage circuitry. Do not allow your fingers to come in contact with the metal probes on the meter.

PROCEDURES	**FINDINGS**	**CONCLUSIONS**
1. Insert the black test lead of the digital multimeter into the "common -" test jack.	—	—
2. Insert the banana plug end of the red test lead into the front panel jack labeled "+".	—	—
3. Place the function switch on a high enough scale to measure 1.5 volts.	—	—
4. Connect the red lead to "+" end of the D battery, and the black lead to "-" end of the battery.	4. Reversing the test leads will result in a "-" reading on the meter.	
5. Observe the reading on the DMM meter face.	5. Record the voltage reading in the blank below. V = _____	5. Compare the readings obtained with the D battery and the C battery._____ What was the difference in the voltage readings? _____.
6. Repeat steps 6 and 7 for the C cell.	6. Record the voltage reading in the blank below. V = _____	6. Why are there several sizes of battery (D, C, AA, AAA), if all have the same voltage rating? _____. What would be the effect if the meter leads were reversed in step 6? _____.

PROCEDURES	FINDINGS	CONCLUSIONS

7. Place the function switch on a scale that will allow reading of at least 9 volts, if the present scale will not handle 9 volts.

—

—

8. Measure the voltage across the terminals of the 9 volt battery.

8. Record the voltage reading in the blank below.

V = _____

8. Why is the voltage from the transistor radio battery so much higher than the D cell, which is larger in physical size?

_____
_____
_____
_____.

Could the C and D cell be connected together to produce 3 volts? Explain.

_____
_____
_____
_____.

Experiment 14

BATTERIES — AIDING AND OPPOSING

SCHEMATIC:

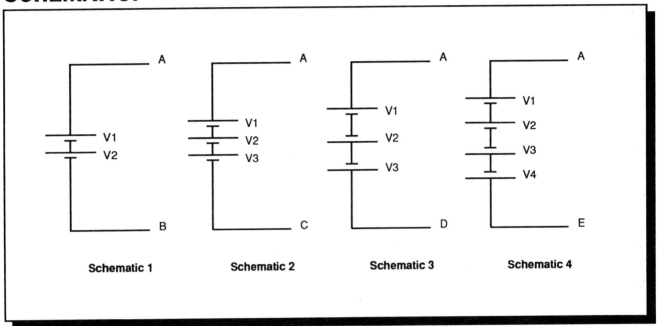

Schematic 1   Schematic 2   Schematic 3   Schematic 4

DISCUSSION

Batteries, as well as electronic power supplies, are sometimes connected in series or in parallel with each other, either to achieve higher voltages than can be obtained with one power supply, or to gain additional current. They must be connected correctly, or, as this lab project will illustrate, they may not produce the results we are trying to obtain.

TEXT CORRELATION

Before beginning this experiment, review section 4–4 on combining voltage in *Electricity*.

OBJECTIVES

1. To connect batteries in series, aiding each other, and measure the results.

2. To connect batteries in series, opposing each other, and measure the results.

EQUIPMENT

- ☐ Standard tool box
- ☐ Digital multimeter, with 10 megohm input resistance on voltage scales
- ☐ 4 assorted batteries, 1.5 volt rating, with hookup wires attached, or installed in a battery holder
- ☐ Standard parts kit
- ☐ Volt-ohm-milliammeter, 20 kilohm/volt DC

PROCEDURES

1. Measure the voltage across the four batteries individually, using the digital multimeter, or the volt-ohm-milliameter, as directed by the instructor.

2. Connect the circuit shown in Schematic 1. Based on the measured values of V_1 and V_2, calculate V_{ab}.

3. Measure V_{ab}.

4. Repeat steps 2 and 3 for the circuit in Schematic 2.

5. Measure V_{ac}.

FINDINGS

1. Record the results below:

   V_1_____ V_2_____
   V_3_____ V_4_____

2. Record the result in the blank below.

   V_{ab} (calculated)_____

3. Record the value of V_{ab} in the blank below.

   V_{ab} (measured)_____

4. Calculate the expected value of V_{ac} and record.

   V_{ac} (calculated)_____

5. Record the value of V_{ac}.

   V_{ac} (measured)_____

CONCLUSIONS

1. State the general rule for adding batteries (or DC power supplies) in series to get more voltage from the combination than from any individual battery

   _____
   _____
   _____
   _____

2. Can you think of an instance where it might be desirable to connect two batteries in series opposing each other?

   _____
   _____
   _____
   _____

3. What would be the output voltage of two 9 volt batteries connected in series (series aiding) with each other?

   _____
   _____
   _____
   _____

4. —

5. Some of the batteries measure considerably larger voltages than the others. Explain why this would be so.

   _____
   _____
   _____
   _____

PROCEDURES	FINDINGS	CONCLUSIONS
6. Repeat steps 2 and 3 for the circuit in Schematic 3.	6. Calculate the expected value of V_{ad} and record. V_{ad} (calculated) _____. Record the measured value of V_{ad}. V_{ad} (measured) _____.	—
7. Repeat steps 2 and 3 for the circuit in Schematic 4.	7. Calculate the expected value of V_{ae}. V_{ae} (calculated) _____. Record the measured value of V_{ae}. V_{ae} (measured) _____.	—

Experiment 15

USING THE ADJUSTABLE DC POWER SUPPLY

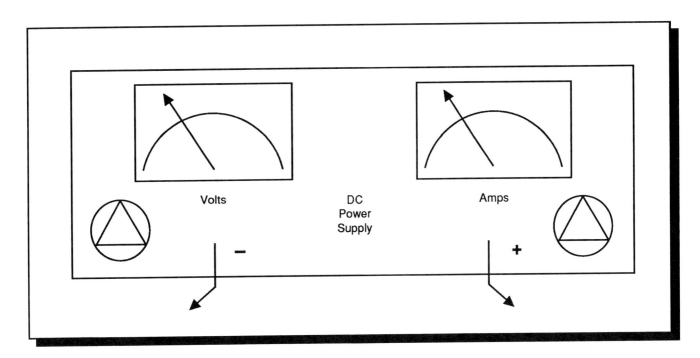

DISCUSSION

The adjustable power supply makes it possible to set any required voltage needed to operate the circuits you will be building in the lab experiments. You will use it for nearly all of the experiments performed in the DC portion of this manual, as well as many of the experiments in the AC portion of the manual, and in later labs working with transistors and ICs.

TEXT CORRELATION

Before beginning this experiment, review section 2–4 on voltage sources in *Electricity*.

OBJECTIVES

1. To perform the necessary procedures for correct operation of the adjustable DC power supply.

2. To measure the output voltage of the DC power supply, at various settings of the function controls on the front panel.

EQUIPMENT

- [ ] Digital multimeter, with 10 megohm input resistance on voltage scales OR
- [ ] Volt-ohm-milliammeter, 20 kilohm/volt DC
- [ ] Adjustable power supply, capable of 0 to 40 VDC output
- [ ] Users manual for the power supply

Using the Adjustable DC Power Supply

SAFETY NOTES

 Even though the output of the DC power supply is below 50 volts, you are developing safety habits that you will use with higher, more dangerous voltages. Exercise caution in taking voltage measurements, to prevent becoming part of the circuit.

PROCEDURES	FINDINGS	CONCLUSIONS
1. Examine the front panel of the adjustable power supply assigned to you.	—	—
2. Using the users manual for the power supply assigned to you, find the operating instructions. Refer to the operating instructions whenever a question arises as you proceed through this procedure.	—	
3. Locate the ON/OFF switch on the front panel. Place it in the ON position.	3. Power supply should be on. If there is a power indicator light, it should be lit.	—
4. Locate the voltage adjust control on the front panel. Rotate the control fully CCW (counterclockwise).	4. This should adjust the power supply for an output of 0 volts.	—
5. Examine the front panel to determine if there is a current limiting control. This is an adjustable control to set the maximum current the supply will allow to flow. Refer to the users manual, if any question arises.	—	—
6. If found, rotate the current limiting control for maximum output of current. (Usually it will be maximum clockwise.)	—	—
7. Locate the test jacks on the front panel. One should be labeled "+" and the other labeled "-". Connect test leads to these two jacks.	7. Note: Many power supplies also have a ground jack on the front panel. If yours does ignore it for now.	7. If your power supply has a front panel meter to indicate voltage output, how does this compare to the actual readings obtained with the voltmeter? _____ _____ _____ _____

Experiment 15   Using the Adjustable DC Power Supply ■ 67

PROCEDURES	FINDINGS	CONCLUSIONS

8. Using either the VOM or the DMM, set up to measure DC voltage, connect the meter to the front panel jacks just described.

8. Note: If using the VOM, it is important that the - lead of the power supply connect to the - lead of the VOM, and the + lead of the power supply connect to the + lead of the VOM.

8. What is the maximum output voltage your power supply is rated for?_____

_____.

Can you obtain this voltage? If not, how close did you come to the maximum?

_____.

_____.

When you adjusted your power supply, did it remain at the voltage you set it at?

_____.

_____.

If not, how much did it change?

_____.

_____.

9. Slowly adjust the voltage adjust control and observe the reading on meter. If the power supply has a voltage meter on the front panel, observe the reading also on the front panel meter.

—

—

10. Adjust the power supply to different voltages, making sure to change the scale on the meter as needed to avoid overloading the meter.

—

—

Experiment 16

OHM'S LAW

SCHEMATIC:

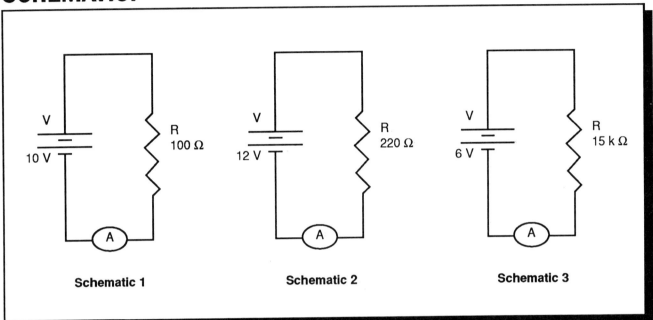

Schematic 1   Schematic 2   Schematic 3

DISCUSSION

Ohm's Law is the single most important concept the technician will ever learn. All troubleshooting, all circuit analysis, all electronic computations are related to Ohm's Law, which simply says that the amount of current that will flow in a circuit will be directly affected by the amount of voltage applied to the circuit, and inversely affected by the amount of resistance in the circuit.

TEXT CORRELATION

Before beginning this experiment, review section 6–6 on Ohm's Law with engineering units in *Electricity*.

OBJECTIVES

1. To measure the amount of current that flows in a circuit for different voltages applied to the circuit.

2. To measure the amount of current that flows in a circuit for different resistance values in the circuit.

3. To verify the concepts involved in Ohm's Law, and the relationship between voltage, resistance, and current.

EQUIPMENT

- ☐ Standard tool box
- ☐ Standard parts kit
- ☐ Adjustable power supply, capable of 0 to 40 VDC output
- ☐ Digital multimeter, with 10 megohm input resistance on voltage scales OR
- ☐ Volt-ohm-milliammeter, 20 kilohm/volt DC

SAFETY NOTES

 Even though the output of the DC power supply is below 50 volts, you are developing safety habits that you will use with higher, more dangerous voltages. Exercise caution in taking voltage measurements, to prevent becoming part of the circuit.

PROCEDURES	FINDINGS	CONCLUSIONS
1. Locate the ON/OFF switch on the front panel of the power supply. Place it in the ON position.	1. Power supply should be on. If there is a power indicator light, it should be lit.	—
2. Locate the voltage adjust control on the front panel. Rotate the control fully CCW (counterclockwise).	2. Power supply output should be 0 volts.	—
3. Connect the meter to the front panel jacks of the power supply, and adjust the power supply to obtain the voltage shown on the schematic diagram. Refer to the users manual, if any question arises.	—	—
4. Connect the circuit shown in Schematic 1.	—	—
5. Calculate the value of current that will flow in the circuit depicted in Schematic 1. Place the function switch on the DMM or VOM on a high enough scale to measure calculated current. SAFETY HINT: It is a good idea to set the ammeter to a higher scale than necessary, until it can be verified that the current is small enough to be safely read on the lower scale.	5. Record the calculated value in the blank below: _____	—
6. Observe the reading on the meter.	6. Record the reading obtained in the blank below: _____	6. In Schematic 1, if the resistor size had been doubled, what would you predict would happen? _____ _____ _____ _____

Experiment 16 Ohm's Law ■ 71

PROCEDURES	**FINDINGS**	**CONCLUSIONS**
7. Locate the ON/OFF switch on the front panel of the power supply. Place it in the OFF position. Disconnect the power supply from the circuit. Using the meter set up as an ohmmeter, measure the resistance of the resistor in the circuit.	7. Record the reading obtained in the blank below: _____	—
8. Connect the circuit shown in Schematic 2.	—	—
9. Repeat the procedure for the circuit in Schematic 2.	9. Record the calculated value in the blank below: _____ Record the reading obtained in the blank below: _____	9. In Schematic 2, what would happen if the voltage is doubled? _____ _____ _____ _____.
10. Repeat the procedure for the circuit in Schematic 3.	10. Record the calculated value in the blank below: _____ Record the reading obtained in the blank below: _____	10. In your own words, state the principle you have discovered in this experiment. _____ _____ _____ _____.

Experiment 17

SIMPLE SERIES CIRCUITS

SCHEMATIC:

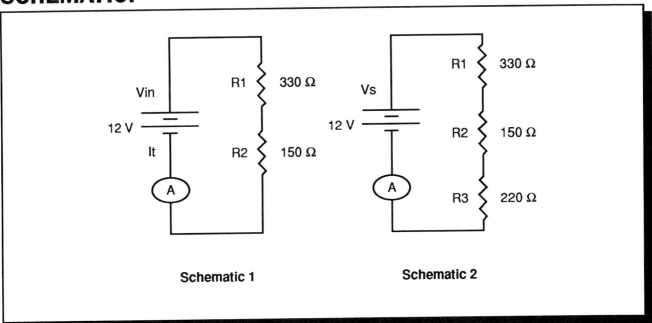

Schematic 1   Schematic 2

DISCUSSION

In electronics, every circuit is connected either as a series circuit, as a parallel circuit, or as a combination series-parallel circuit. The series circuit operates under a different set of rules than the parallel circuit. It is necessary for the technician to be able to analyze series circuits, so that he/she can troubleshoot and make adjustments to the circuit.

TEXT CORRELATION

Before beginning this experiment, review section 9–1 on current and voltage in series in *Electricity*.

OBJECTIVES

1. To connect components in a series circuit.

2. To analyze the series circuit and determine what physics principles influence the operation of the series circuit.

EQUIPMENT

- ☐ Standard tool box
- ☐ Standard parts kit
- ☐ Volt-ohm-milliammeter, 20 kilohm/volt DC
- ☐ Digital multimeter, with 10 megohm input resistance on voltage scales
- ☐ Adjustable power supply, capable of 0 to 40 VDC output

PROCEDURES | FINDINGS | CONCLUSIONS

1. Locate the ON/OFF switch on the front panel of the power supply. Place it in the OFF position.

    —

    —

2. Locate the voltage adjust control on the front panel. Rotate the control fully CCW (counterclockwise). This should adjust the power supply for an output of 0 volts.

    —

    —

3. Calculate the value of voltage drop across each of the resistors, and enter in the blanks.

    3. Record the calculated values in the blanks below:

    V_{R1} ——————

    V_{R2} ——————

    3. What statement can you make concerning total resistance as resistors are added in series?

    ——————
    ——————
    ——————
    ——————

4. Connect the circuit shown in Schematic 1.

    —

    —

5. Locate the test jacks on the front panel of the power supply. One should be labeled "+" and the other labeled "-". Connect test leads to these two jacks. (Note: Many power supplies also have a ground jack on the front panel. If so, ignore it for now.)

    —

    —

6. Using either the VOM or the DMM, set up to measure DC voltage, connect the meter to the front panel jacks just described. (Note: If using the VOM, it is important that the - lead of the power supply connect to the - lead of the VOM, and the + lead of the power supply connect to the + lead of the VOM.)

    —

    —

7. Energize the power supply by placing the ON/OFF switch in the ON position.

    —

    —

PROCEDURES	FINDINGS	CONCLUSIONS
8. Slowly adjust the voltage adjust control and observe the reading on meter. If the power supply has a voltage meter on the front panel, observe the reading also on the front panel meter. Adjust the power supply to the voltage indicated on Schematic 1.	—	—
9. Measure the voltage drops across each of the resistors in the circuit. Record the results.	9. Record the readings obtained in the blanks below: V_{R1} ───────── V_{R2} ─────────	9. Describe, in your own words, how the voltages are distributed around the circuit. What effect does the relative size of the resistors have concerning voltage drops? ───────── ───────── ───────── ─────────.
10. Locate the ON/OFF switch on the front panel of the power supply. Place it in the OFF position.	—	—
11. Remove the meter from the circuit.	—	—
12. Calculate the value of current that will flow in the circuit depicted in Schematic 1. Place the function switch on the meter on a high enough scale to measure calculated current. SAFETY HINT: It is a good idea to set the ammeter to a higher scale than necessary, until it can be verified that the current is small enough to be safely read on the lower scale	12. Record the calculated value in the blank below: I_t ───────── ─────────.	12. How is current related to the voltage applied to the series circuit? ───────── ───────── ───────── ─────────.
13. Install the multimeter, set up as an ammeter, in series with the circuit. Reenergize the power supply. Observe readings of the current through the circuit, and enter.	13. Record the reading obtained in the blank below: I_t ───────── ─────────.	—

76 ■ Simple Series Circuits

PROCEDURES | FINDINGS | CONCLUSIONS

14. Deenergize the power supply. Disconnect the power supply leads from the circuit, and measure the total resistance of the circuit, as well as the resistance of each individual resistor. Enter the information in blanks.

15. Repeat steps 1 through 14 for the circuit in Schematic 2. Enter the appropriate information in the blanks.

14. Record the readings obtained in the blanks below:

R_t ─────────
R_1 ─────────
R_2 ─────────

15. Record the calculated values in the blanks below:

V_{R1} ─────────
V_{R2} ─────────

Record the readings obtained in the blanks below:

V_{R1} ─────────
V_{R2} ─────────

Record the calculated value in the blank below:

I_t ─────────

Record the reading obtained in the blank below:

I_t ─────────

Record the readings obtained in the blanks below:

R_t ─────────
R_1 ─────────
R_2 ─────────
R_3 ─────────

14. How is current related to the resistance of the series circuit?

─────────
─────────
─────────
─────────
─────────.

─

Experiment 18

MEASURING RESISTANCE IN SERIES CIRCUITS

SCHEMATIC:

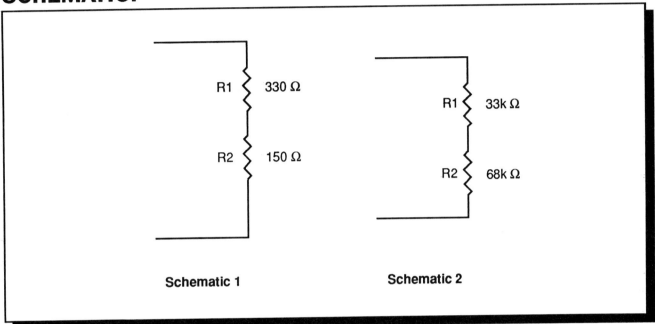

Schematic 1

Schematic 2

DISCUSSION

The resistance in a series circuit can be measured directly, without disconnecting components to make the measurement. This makes it much simpler to measure than parallel circuits. The total resistance in a series circuit is always simply the sum of each of the individual resistors. So, if a circuit consists of two resistors connected in series, all that it is necessary to do to determine the total effect on the circuit is to add the two values together. For three resistors, add the three values together.

TEXT CORRELATION

Before beginning this experiment, review section 9–2 on Ohm's Law applied to series circuits in *Electricity*.

OBJECTIVES

1. To measure the resistance of each resistor in the circuit in a series connected circuit.

2. To measure the total resistance of the circuit in a series connected circuit.

EQUIPMENT

☐ Standard tool box

☐ Standard parts kit

☐ Digital multimeter, with 10 megohm input resistance on voltage scales OR

☐ Volt-ohm-milliammeter, 20 kilohm/volt DC

Measuring Resistance in Series Circuits

Experiment 18

PROCEDURES	FINDINGS	CONCLUSIONS
1. Using the resistor color code, select the 47 ohm and 10 ohm resistors from the standard parts kit.	—	—
2. Connect the circuit shown in Schematic 1.	—	—
3. Calculate the expected total resistance of the circuit. Record this value.	3. Record the calculated value in the blank below: _____ _____	3. State the relationship of individual resistors to the total resistance; in other words, how does the addition or subtraction of a resistor to the circuit affect the total resistance of the circuit? _____ _____ _____ _____
4. Set up the meter to read this expected value of resistance.	—	—
5. Connect the meter to the circuit and measure total resistance of the circuit. Record this value.	5. Record the reading obtained in the blank below: _____ _____	5. The measured value of resistance probably didn't match up exactly with the calculated value. Can you explain why this happened? _____ _____ _____ _____
6. Determine the expected resistance reading of each individual resistor, if still connected in the circuit. Record this value.	6. Record the calculated values in the blanks below: R_1 _____ R_2 _____	—
7. Connect the meter across each resistor in turn, measuring the resistance of each resistor. Record these readings.	7. Record the readings obtained in the blanks below: R_1 _____ R_2 _____	7. How close did the measured values of the resistors come to the nominal value for the resistors? _____ Is this within acceptable tolerance? _____ _____ _____

PROCEDURES	FINDINGS	CONCLUSIONS
8. Connect the circuit as in Schematic 2. Repeat the above procedure for the new circuit, recording the information in the blank.	8. R_1 _____ R_2 _____ R_t _____	8. How close did the measured values of the resistors come to the nominal value for the resistors? _____ Is this within acceptable tolerance? _____ _____ _____ _____.
9. Connect a test lead (piece of hookup wire or an alligator clip test lead) across R_1 to simulate a shorted resistor.	—	—
10. Determine the expected total resistance, as well as the resistance reading across each of the resistors individually, and record in the blank.	10. R_1 _____ R_2 _____ R_t _____	—
11. Measure the total resistance, and the individual resistances, and record in the blank.	11. R_1 _____ R_2 _____ R_t _____	11. How close did the measured values of the resistors come to the nominal value for the resistors? _____ Is this within acceptable tolerance? _____ _____ _____ _____.
12. Remove the short circuit from R_1. Disconnect R_2 by removing the jumper between R_1 and R_2. This will simulate an open resistor.	—	—
13. Determine the expected total resistance, as well as the resistance reading across each of the resistors individually, and record in the blank.	13. R_1 _____ R_2 _____ R_t _____	—

80 ■ Measuring Resistance in Series Circuits Experiment 18

| **PROCEDURES** | **FINDINGS** | **CONCLUSIONS** |

14. Measure the total resistance, and the individual resistances, and record in blank.

14. R_1 _____
    R_2 _____
    R_t _____

14. How close did the measured values of the resistors come to the nominal value for the resistors? _____ Is this within acceptable tolerance?

_____
_____
_____
_____

Experiment 19

VOLTAGE MEASUREMENTS IN SERIES CIRCUITS

SCHEMATIC:

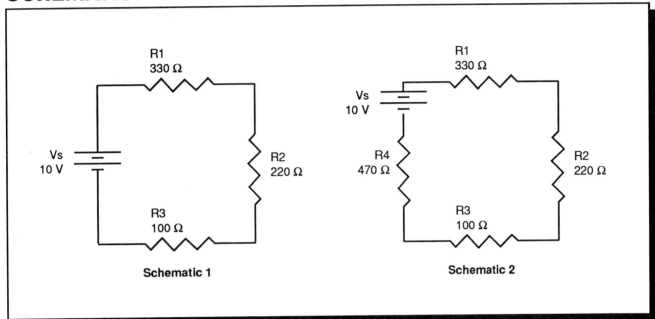

Schematic 1

Schematic 2

DISCUSSION

One of the necessary skills the new technician must develop is the ability to measure voltages around a circuit. This is necessary in order to determine whether the circuit is operating correctly, and, if not, what is wrong with the operation of the circuit. In series circuits, the sum of all the individual voltage readings around the circuit must add up to the total voltage applied to the circuit.

TEXT CORRELATION

Before beginning this experiment, review section 9–3 on Kirchoff's Voltage Law in *Electricity*.

OBJECTIVES

1. To measure the voltages around a series circuit.
2. To determine whether the voltages are correct.
3. To measure the effect on voltage readings of a short circuit and an open circuit.

EQUIPMENT

- ☐ Standard tool box
- ☐ Standard parts kit
- ☐ Volt-ohm-milliammeter, 20 kilohm/volt DC
- ☐ Digital multimeter, with 10 megohm input resistance on voltage scales
- ☐ Adjustable power supply, capable of 0 to 40 VDC output

SAFETY NOTES

Even though the voltage across the resistors is very low, observe all safety precautions for higher voltage circuits. Remember that you are developing habits with low voltage circuits that you will use with higher voltage circuits.

PROCEDURES	FINDINGS	CONCLUSIONS
1. Locate the ON/OFF switch on the front panel of the power supply. Place it in the OFF position.	1. The power supply should be OFF.	—
2. Locate the voltage adjust control on the front panel. Rotate the control fully CCW (counterclockwise).	2. The output of the power supply should be 0 volts.	—
3. Set up the multimeter to read DC volts. Connect the meter to the front panel jacks marked "-" and "+". Reenergize the power supply. Slowly adjust the voltage adjust control and observe the reading on the meter.	3. Adjust the power supply to the voltage indicated on Schematic 1.	—
4. Deenergize the power supply. Connect the circuit shown in Schematic 1. Calculate all voltage drops, total resistance, and current flow.	4. Record the calculated voltage drops, total resistance, and current flow in the blanks below. V_{R1} ——— V_{R2} ——— V_{R3} ——— R_t ——— I_t ———	4. What statement can you make about the calculated voltages around the series circuit compared to the voltage applied? ————— ————— ————— —————
5. Reenergize the power supply. Observe voltage readings across the resistors. Deenergize the power supply. Disconnect the power supply leads from the circuit, and measure the total resistance of the circuit. (Note: If using the VOM, it is important that the - end of the resistor connect to the - lead of the VOM, and the + lead of the resistor connect to the + lead of the VOM.)		If R_1 were increased in size in Schematic 1, how would this affect the voltage measured across it? ————— ————— ————— —————

Experiment 19 Voltage Measurements in Series Circuits ■ 83

PROCEDURES	FINDINGS	CONCLUSIONS
—	5. Observe voltage readings across the resistors and the total resistance, and enter in the blanks below. V_{R1} _____ V_{R2} _____ V_{R3} _____ R_t _____	5. What statement can be made about the measured voltages around the series circuit compared to the calculated voltages? _____ This would tend to prove that voltages around a series circuit must _____ _____ (add up to/subtract from) the total voltage applied to the circuit.
6. Deenergize the power supply. Connect the circuit shown in Schematic 2. Calculate all voltage drops, total resistance, and current flow for this new circuit.	6. Record the calculated values for the circuit in Schematic 2 in the blanks below. V_{R1} _____ V_{R2} _____ V_{R3} _____ V_{R4} _____ R_t _____ I_t _____	—
7. Reenergize the power supply. Adjust the power supply as necessary to correspond with the indicated voltage on Schematic 2. Observe readings of the voltages across the resistors, and record them.	7. Record the measured values for the circuit in Schematic 2 in the blanks below. V_{R1} _____ V_{R2} _____ V_{R3} _____ V_{R4} _____	—
8. Deenergize the power supply. Disconnect the leads from the circuit, and measure the total resistance of the circuit. Record the information.	8. Record the value measured for R_t. R_t _____	8. If an extra resistor of 2.2 kilohm is added, in series, to the circuit in Schematic 2, how will this affect the voltage drops across each of the resistors in the circuit? _____ _____ _____

84 ■ Voltage Measurements in Series Circuits   Experiment 19

PROCEDURES   FINDINGS   CONCLUSIONS

How about the total resistance?

_____

_____

_____

_____

Experiment 20

MEASURING CURRENT IN SERIES CIRCUITS

SCHEMATIC:

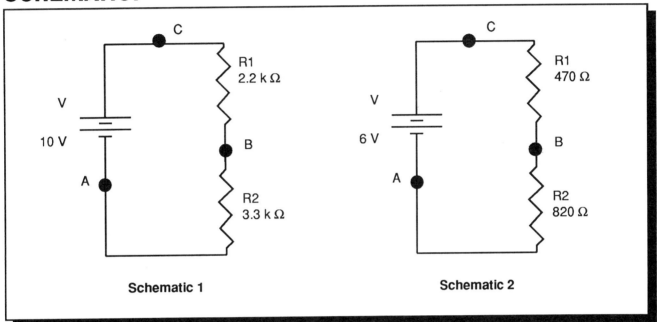

Schematic 1   Schematic 2

DISCUSSION

The current that flows in a series circuit should be the same at any point in the circuit. In other words, no matter where in the circuit the ammeter is inserted, the current recorded should be identical. The technician needs to know this information, since he/she will be required to measure the current in the circuit and use this information to determine whether the circuit is operating correctly or not.

TEXT CORRELATION

Before beginning this experiment, review section 9–1 on current and voltage in series in *Electricity*.

OBJECTIVES

1. To measure the current that flows in a series circuit, using either a VOM or DMM meter.

EQUIPMENT

- ☐ Standard tool box
- ☐ Standard parts kit
- ☐ Adjustable power supply, capable of 0 to 40 VDC output
- ☐ Digital multimeter, with 10 megohm input resistance on voltage scales OR
- ☐ Volt-ohm-milliammeter, 20 kilohm/volt DC

SAFETY NOTES

Be absolutely certain before energizing the power supply, that the ammeter is connected in series with the circuit to be tested. Ammeters should never be connected in parallel with a component. To do so will cause the component to be short circuited through the ammeter, and may damage the meter or circuit, or the power supply. If you have any doubt, check with your instructor.

PROCEDURES	FINDINGS	CONCLUSIONS
1. Locate the ON/OFF switch on the front panel of the power supply. Place it in the OFF position.	—	—
2. Locate the voltage adjust control on the front panel. Rotate the control fully CCW (counterclockwise).	—	—
3. Locate the ON/OFF switch on the front panel of the power supply. Place it in the ON position. Slowly adjust the voltage adjust control until the output is the same as the voltage indicated in Schematic 1.	—	—
4. Connect the circuit shown in Schematic 1. Calculate total resistance and current flow.	4. Record the calculated values in the blanks below: R_t _____ I_t _____	
5. Set up the multimeter (either the VOM or the DMM) to measure DC Current. Place the meter on a high enough scale to ensure safe reading of the amount of current you have calculated, and connect the ammeter to point A. SAFETY HINT: Since the meter can be seriously damaged by improperly measuring current, always place the ammeter on the highest scale available for the initial reading, then switch to a lower scale only after noting that the initial reading will "fit" that scale.	—	5. The multimeter, when used as an ammeter, has a very low resistance (often less than .5 ohms). What special precautions must you observe because of this? _____ _____ _____ _____.

PROCEDURES	FINDINGS	CONCLUSIONS

— | 6. Record the reading obtained in the blank below:

I_t ———————— | —

7. Deenergize the power supply. Disconnect the leads from the circuit, and move the ammeter to the second point (point B) indicated on the schematic diagram. Reenergize the circuit, and, once again, observe the reading on the ammeter. | 7. Record the reading obtained in the blank below:

I_t ———————— | —

8. Repeat step 7 for the third point (point C) indicated on the schematic diagram. Record the result. | 8. Record the reading obtained in the blank below:

I_t ———————— | —

9. Repeat steps 3 through 8 for the circuit shown in Schematic 2. Record all results. | 9. Record the calculated values in the blanks below:

R_t ————————

I_t ————————

Record the reading obtained in the blank below:

I_t ———————— | 9. What conclusion can you arrive at, based on what you learned in this experiment, concerning current measurements taken at several different points in a series circuit?

——————————
——————————
——————————
——————————.

10. Remove the meter from the circuit. | — | 10. In actual practice, technicians do not measure current very often. Instead, they indirectly measure current, by measuring voltage and resistance and calculating the current from these readings. What two reasons can you think of why this might be so?

——————————
——————————
——————————
——————————.

Experiment 21

CONNECTING SERIES CIRCUITS USING GROUND

SCHEMATIC:

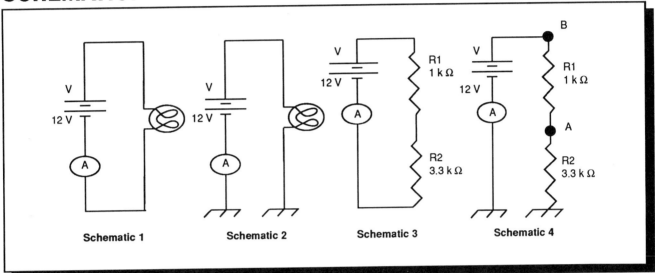

Schematic 1   Schematic 2   Schematic 3   Schematic 4

DISCUSSION

The concept of ground is an important one for the electronics technician, since most equipment is connected such that ground is used as one of the paths for current to flow in the circuitry. Whether the circuit is actually connected to the earth (earth ground), or is connected to a conductive metal chassis (chassis ground), the effect is the same. The ground connection acts just like a wire to conduct current. It is also used as a common point from which to measure nearly all voltages and resistances.

TEXT CORRELATION

Before beginning this experiment, review section 9–5 on potentiometer voltage dividers in *Electricity*.

OBJECTIVES

1. To connect series circuits such that ground is used as a return line for the current to flow.

2. To measure voltages and resistances in respect to ground.

EQUIPMENT

- ☐ Standard tool box
- ☐ Standard parts kit
- ☐ Digital multimeter, with 10 megohm input resistance on voltage scales OR
- ☐ Volt-ohm-milliammeter, 20 kilohm/volt DC
- ☐ Adjustable power supply, capable of 0 to 40 VDC output
- ☐ One 12 volt automotive light bulb, type #1004 or #1157 or equivalent
- ☐ Metal rod or metal chassis to be used as chassis ground

SAFETY NOTES

 Be absolutely certain before energizing the power supply, that the ammeter is connected in series with the circuit to be tested. Ammeters should never be connected in parallel with a component. To do so will cause the component to be short circuited through the ammeter, and may damage the meter or circuit, or the power supply. If you have any doubt, check with your instructor.

PROCEDURES	FINDINGS	CONCLUSIONS
1. Locate the ON/OFF switch on the front panel of the power supply. Place it in the OFF position.	1. The power supply should be OFF. The power indicator light or LED should be OFF.	1. In your own words, explain the principles involved in measuring voltages in respect to ground. _____ _____ _____ _____ Why do we prefer to take measurements in respect to ground, rather than across each component? _____ _____ _____ _____
2. Locate the voltage adjust control on the front panel. Rotate the control fully CCW (counterclockwise).	—	—
3. Connect the circuit shown in Schematic 1.	—	—
4. Energize the circuit, by placing the ON/OFF switch in the ON position. Measure the current in the circuit in Schematic 1.	4. Note: When measuring the current in the circuit, put the ammeter on its highest scale. The current drain with the automotive bulb may be greater than .5 amperes. Record the reading obtained in the blank below: I (measured) _____	4. Discuss the safety considerations involved in taking measurements in respect to ground rather than across individual components. _____ _____ _____ _____
5. Place the ON/OFF switch on the front panel of the power supply in the OFF position.	—	—

Experiment 21   Connecting Series Circuits Using Ground ■ 91

PROCEDURES	FINDINGS	CONCLUSIONS
6. Locate the voltage adjust control on the front panel. Rotate the control fully CCW (counterclockwise).	6. The power supply should be OFF, and the output should be 0 volts.	—
7. Connect the circuit shown in Schematic 2. Use the metal rod listed in the "Materials" section of this experiment as a ground. Simply connect the negative end of the voltage source to one end of the metal rod, and one end of the lamp — the end that is not connected to the switch — to the other end of the metal rod. (Note: The metal rod is not actually connected to earth ground, so the schematic diagram uses the chassis ground schematic symbol.)	—	—
8. Measure the current that flows in the circuit of Schematic 2.	8. Record the reading obtained in the blank below: I (measured) _____	8. Compare the current readings from Schematic 1 and Schematic 2. What conclusions can you draw from this information? _____ _____ _____ _____.
9. Place the ON/OFF switch on the front panel of the power supply in the OFF position.	—	—
10. Rotate the voltage adjust control on the front panel fully CCW (counter clock wise).	—	—
11. Connect the circuit shown in Schematic 3. Connect a multimeter across the terminals of the power supply.	—	—
12. Switch the ON/OFF switch to the ON position.	—	—

92 ■ Connecting Series Circuits Using Ground

PROCEDURES	FINDINGS	CONCLUSIONS

13. Slowly adjust the voltage adjust control, while observing the reading on the voltmeter, until the value is reached that is indicated on Schematic 3.

—

—

14. Measure all values of current and voltage specified. Record the values.

14. Record the readings obtained in the blanks below:

I (measured)_____

V_{R1} _____

V_{R2} _____

14. What conclusions can you arrive at concerning the magnitude of the voltage drop across the larger resistors in a series circuit compared to the smaller resistors?

_____
_____
_____
_____

15. Place the ON/OFF switch on the front panel of the power supply to the OFF position.

—

—

16. Connect the circuit shown in Schematic 4. Remember to use the metal rod or metal chassis for the chassis ground connection.

—

—

17. Reenergize the power supply by placing the ON/OFF switch in the ON position.

—

—

18. Measure all values of current and voltage specified. Record them in designated place.

18. Record the readings obtained in the blanks below:

I (measured)_____

V_{R1} _____

V_{R2} _____

—

19. Connect the negative lead of the multimeter to the ground (metal rod or metal chassis). Connect the positive lead to the junction between R_1 and R_2. Measure and record the value of V_A.

19. Record the reading obtained in the blank below:

V_A _____

19. Compare the voltage V_A to the voltage V_{R2} measured previously

_____
_____
_____
_____

Experiment 21 — Connecting Series Circuits Using Ground ■ 93

PROCEDURES	FINDINGS	CONCLUSIONS
20. Leave the negative lead of the multimeter connected to ground. Move the positive lead to the other end of R_1 (designated as B on Schematic 4).	20. Record the reading obtained in the blank below: V_B _____	20. Compare the voltage V_B to the voltage set on the power supply. _____ _____ _____ _____

Experiment 22

VOLTAGE DIVIDERS (UNLOADED)

SCHEMATIC:

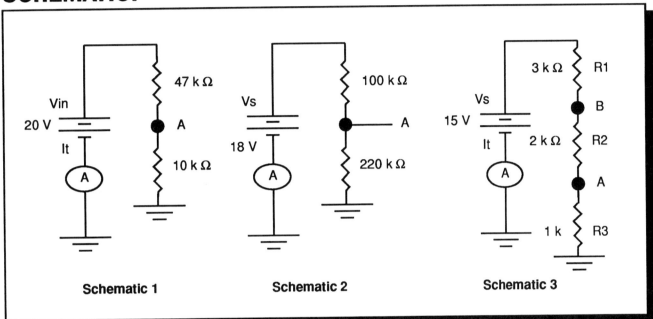

DISCUSSION

Voltage dividers are used in many applications in electronics, for providing multiple voltages from a single power supply. The technician must learn to analyze voltage dividers, so he/she can determine whether the circuit is operating correctly or not, and, if the circuit is not operating correctly, to determine what is necessary to effect repairs.

TEXT CORRELATION

Before beginning this experiment, review section 9–2 on Ohm's Law Applied to Series Circuits in *Electricity*.

OBJECTIVES

1. To calculate the values associated with voltage divider circuits.

2. To construct voltage dividers and analyze the operation of the circuit.

EQUIPMENT

- ☐ Standard tool box
- ☐ Standard parts kit
- ☐ Digital multimeter, with 10 megohm input resistance on voltage scales
- ☐ Volt-ohm-milliammeter, 20 kilohm/volt DC
- ☐ Adjustable power supply, capable of 0 to 40 VDC output

95

96 ■ Voltage Dividers (Unloaded)   Experiment 22

PROCEDURES	FINDINGS	CONCLUSIONS

1. Place the ON/OFF switch on the front panel of the power supply in the OFF position.

2. Rotate the voltage adjust control on the front panel of the power supply fully CCW (counterclockwise). This will set the output of the power supply to zero volts.

3. Connect the circuit shown in Schematic 1.

                                            3. What will a circuit like this be used for?

                                            _____
                                            _____
                                            _____
                                            _____.

4. Set up the meter to measure DC volts, and connect it across the output terminals of the power supply.

                4. If using a VOM, be certain that the "–" lead of the meter is connected to the "–" lead of the power supply, and the "+" lead is connected to the "+" lead of the power supply.

5. Reenergize the power supply by placing the ON/OFF switch in the ON position. Adjust the voltage adjust control until the value measured on the voltmeter equals the value indicated on the schematic diagram.

6. Measure all values specified, and record them.

                6. Record the readings obtained in the blanks below:

                    V_{R1} _____

                    V_{R2} _____

                    V_A _____

                                            6. If you have 4 different resistors, how many voltages can you get from one power supply, including the power supply voltage?

                                            _____

                                            Why? _____

                                            _____

                                            _____.

Experiment 22   Voltage Dividers (Unloaded) ■ 97

PROCEDURES	FINDINGS	CONCLUSIONS
7. Place the ON/OFF switch on the front panel of the power supply to the OFF position.	—	7. If there were a short across R_2 in Schematic 2, how would this affect the voltage output at A? _____.
8. Rotate the voltage adjust control on the front panel of the power supply fully CCW (counterclockwise). This will set the output of the power supply to zero volts.	—	—
9. Connect the circuit shown in Schematic 2.	—	9. If R_1, in Schematic 2, were open, how would this affect the output voltage at A? _____. Why?_____.
10. Reenergize the power supply by placing the ON/OFF switch in the ON position. Adjust the voltage adjust control until the value measured on the voltmeter equals the value indicated on the schematic diagram.	—	—
11. Measure all values specified, and record them.	11. Record the readings obtained in the blanks below: V_{R1}_____ V_{R2}_____ V_A_____	—

PROCEDURES	FINDINGS	CONCLUSIONS
12. Place the ON/OFF switch on the front panel of the power supply to the OFF position.	—	—
13. Rotate the voltage adjust control on the front panel of the power supply fully CCW (counterclockwise). This will set the output of the power supply to zero volts.	—	—
14. Connect the circuit shown in Schematic 3.	—	—
15. Reenergize the power supply, by placing the ON/OFF switch in ON position. Adjust the voltage adjust control until the value measured on the voltmeter equals the value indicated on the schematic diagram.	—	—
16. Measure all values specified, and record them.	16. Record the readings obtained in the blanks below: V_{R1} —————— V_{R2} —————— V_{R3} —————— V_A —————— V_B ——————	—

Experiment 23

SERIES CIRCUITS DESIGN CONSIDERATIONS

SCHEMATIC:

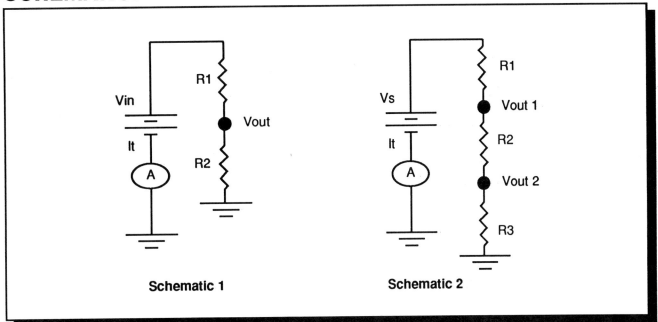

DISCUSSION

Often the technician will need to design a circuit or a portion of a circuit to accomplish a particular job. For example, it may be necessary to design a voltage divider circuit to supply a smaller voltage to a circuit than is readily available from the power supply on hand. The technician needs to know how to select the necessary resistors.

TEXT CORRELATION

Before beginning this experiment, review section 9–2 on Ohm's Law applied to series circuits in *Electricity*.

OBJECTIVES

1. To select the resistors necessary to design a circuit to produce a specific voltage, from a power supply rated at a higher voltage.
2. To assemble the circuit and test it do determine whether the circuit is operating properly.

EQUIPMENT

- ☐ Standard tool box
- ☐ Standard parts kit
- ☐ Adjustable power supply, capable of 0 to 40 VDC output
- ☐ Digital multimeter, with 10 megohm input resistance on voltage scales OR
- ☐ Volt-ohm-milliammeter, 20 kilohm/volt DC

PROCEDURES | FINDINGS | CONCLUSIONS

Design Problem 1

1. Using two resistors from the standard parts kit, design a series circuit to have the following specifications. (Refer to Schematic 1.)

    V_{out} = 8.5 volts

    I_{total} less than 700 µA

    V_{in} = 24 volts

2. Record the values, including wattage rating.

    2. List the two resistor values below.

    R_1 = _____

    @ _____ watts

    R_2 = _____

    @ _____ watts

    2. In any good design project, the maximum current through the circuit, and/or the maximum power dissipation, would be an important consideration. Why is this so?

    _____

3. Calculate the output voltage and total current. Enter the values in the blanks.

    3. Record the calculated values in the blanks below:

    V_{out} = _____

    I_{total} = _____

4. Connect the circuit. Measure the output voltage, and the total current, and enter in the blanks.

    4. Record the readings obtained in the blanks below:

    V_{out} _____

    I_t _____

Design Problem 2

5. Using three resistors from the standard parts kit, design a circuit with the following specifications. (Refer to Schematic 2.)

    V_{out1} = 12 volts

    V_{out2} = 6 volts

    V_{in} = 20 volts

    I_{total} = less than 1.3 mA

    5. Why would a technician want to design a circuit that will output several different voltages from the same power supply?

    _____

Experiment 23   Series Circuits Design Considerations ■ 101

PROCEDURES	FINDINGS	CONCLUSIONS
6. List the resistor values to be used in the blanks.	6. Record the calculated values in the blanks below: $R_1 =$ _____ $R_2 =$ _____ $R_3 =$ _____	—
7. Calculate V_{out} and I_{total}. Enter the values in the blanks.	7. Record the calculated values in the blanks below: V_{out1} _____ V_{out2} _____ I_t _____	—
8. Connect the circuit. Measure the output voltage and the total current, and enter in the blanks.	8. Record the readings obtained in the blanks below: V_{out1} _____ V_{out2} _____ I_{total} _____	8. A circuit consisting of a voltage source of 50 volts and two resistors connected in series with each other (see Schematic 3) is needed that will deliver 27 volts to another circuit that is to be connected across one of the resistors. Can you determine the value of the two resistors, if the resistors are to be no larger than 1/2 watt? (Note: Assume the resistance of the load circuit is large enough to be ignored in the calculations.) _____ _____ _____ _____.

Experiment 24

TROUBLESHOOTING SERIES CIRCUITS USING RESISTANCE READINGS

SCHEMATIC:

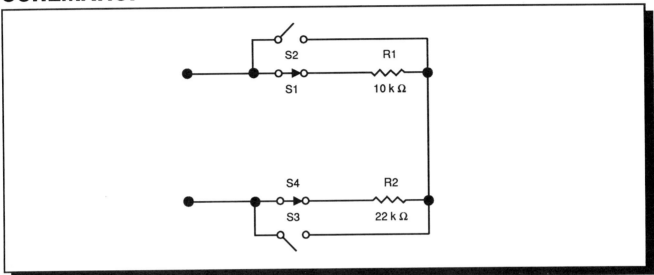

DISCUSSION

Troubleshooting is part of the technician's daily routine. No matter how long he/she is employed in the electronics field, he/she will still be likely to be involved in troubleshooting. Even if the main job becomes design work, which many technicians do, troubleshooting is still a big part of the daily routine. Troubleshooting is simply the science of analyzing the test readings obtained in a circuit and determining from these test readings what may be causing the malfunction. Interpretation of the readings is the key, not the accuracy of the readings. Knowing what the readings mean is all-important. One of the most important readings is the resistance reading. From it the technician can determine whether a circuit is shorted or open, or whether a resistor is out of tolerance.

TEXT CORRELATION

Before beginning this experiment, review section 9–6 on troubleshooting series circuits in **Electricity**.

OBJECTIVES

1. To troubleshoot a series circuit, using resistance readings only, to determine what is malfunctioning in the circuit.

EQUIPMENT

- ☐ Standard tool box
- ☐ Standard parts kit
- ☐ 4 single pole, single throw switches
- ☐ Digital multimeter, with 10 megohm input resistance on voltage scales OR
- ☐ Volt-ohm-milliammeter, 20 kilohm/volt DC

104 ■ Troubleshooting Series Circuits Using Resistance Readings

Experiment 24

SAFETY NOTES

 Notice that there is no power supply shown on the schematics for this lab project. The ohmmeter is never used in a circuit with power applied. To do so can damage the meter.

PROCEDURES	**FINDINGS**	**CONCLUSIONS**
1. Connect the circuit in Figure 1. Place switches in the positions shown on the schematic. (S_1, S_4 closed, S_2, S_3 open)	—	—
2. Measure the resistance readings of each individual resistor, by placing the meter leads (with the meter set up to measure resistance) from point A to point C to measure R_1, and point B to point C to measure R_2. Record the readings in the blanks provided.	2. Record the readings obtained in the blanks below: R_1 _____ R_2 _____	—
3. Measure the total resistance of the circuit by connecting the meter leads across point A to point B. Record the readings.	3. Record the readings obtained in the blanks below: R_t _____	—
4. Close switch S_2. Again, measure the resistance across each individual resistor, by measuring from point A to point C to measure R_1, and point B to point C to measure R_2, and record the results.	4. Record the readings obtained in the blanks below: R_1 _____ R_2 _____	4. Notice what happened to the resistance of R_1. The resistance of R_1 is now _____ ohms, which indicates that the resistor is _____ (shorted/open). What happened to the resistance across R_2? _____ _____ _____ _____.
5. Measure the total resistance, by measuring from point A to point B, and record in the blank indicated.	5. Record the reading obtained in the blank below: R_t _____	5. Notice what happened to the total resistance of the circuit. The total resistance has now _____ (increased/decreased) to _____ ohms.
6. Open Switch S_2, and close Switch S_3.	—	—

Experiment 24 Troubleshooting Series Circuits Using Resistance Readings ■ 105

PROCEDURES	FINDINGS	CONCLUSIONS
7. Measure and record the resistance of the individual resistors, by measuring from point A to point C to measure R_1, and point B to point C to measure R_2.	7. Record the readings obtained in the blanks below: R_1 _____ R_2 _____	7. Notice what happened to the resistance of R_2. The resistance of R_2 is now _____ ohms, which indicates that the resistor is _____ (shorted/open). What happened to the resistance across R_1? _____ _____ _____ _____.
8. Measure and record the total resistance of the circuit, by measuring from points A to B.	8. Record the reading obtained in the blank below: R_t _____	8. Notice what happened to the total resistance of the circuit. The total resistance has now _____ (increased/decreased) to ____ _____ ohms.
9. Open S_1 and S_3. Again, measure and record the resistance of each individual resistor, by measuring from points A to C, and points B to C, as in step 2.	9. Record the readings obtained in the blanks below: R_1 _____ R_2 _____	9. Notice what happened to the resistance of R_1. The resistance of R_1 is now _____ ohms, which indicates that the resistor is _____ (shorted/open). What happened to the resistance across R_2? _____ _____ _____ _____.
10. Measure and record the total resistance, as before, by measuring from point A to point B.	10. Record the reading obtained in the blank below: R_t _____	10. Notice what happened to the total resistance of the circuit. The total resistance has now _____ (increased/decreased) to ____ _____ ohms.
11. Close S_1, and open S_4. Again, measure and record the resistance of R_1 and R_2 by measuring between points A and C, and points B and C.	11. Record the readings obtained in the blanks below: R_1 _____ R_2 _____	11. Notice what happened to the resistance of R_2. The resistance of R_2 is now _____ ohms, which indicates that the resistor is _____

PROCEDURES	FINDINGS	CONCLUSIONS
		(shorted/open). What happened to the resistance across R_1? _____ _____ _____ _____
12. Measure and record the total resistance in the circuit, as in step 10.	12. Record the reading obtained in the blank below: R_t _____	12. Notice what happened to the total resistance of the circuit. The total resistance has now _____ (increased/decreased) to _____ ohms.

Experiment 25

TROUBLESHOOTING SERIES CIRCUITS USING VOLTAGE READINGS

SCHEMATIC:

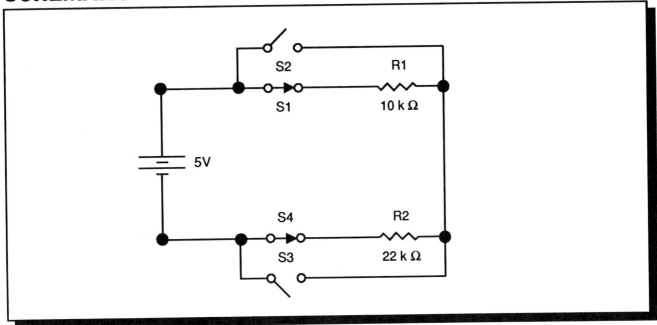

DISCUSSION

One of the most informative tests the technician can perform on an electronic circuit is the measurement of the voltages measured at various points in the circuit. An open circuit, or a short circuit, can be readily identified using voltage readings AND a correct interpretation of what the voltage readings mean. Although more dangerous to perform than resistance readings, voltage readings are, in many cases, faster to perform, and are more informative than resistance readings.

TEXT CORRELATION

Before beginning this experiment, review section 9–6 on troubleshooting series circuits in *Electricity*.

OBJECTIVES:

1. To troubleshoot a series circuit, using voltage readings only, to determine what is malfunctioning in the circuit.

EQUIPMENT

- ☐ Standard tool box
- ☐ Standard parts kit
- ☐ Adjustable power supply, capable of 0 to 40 VDC output
- ☐ Digital multimeter, with 10 megohm input resistance on voltage scales OR
- ☐ Volt-ohm-milliammeter, 20 kilohm/volt DC

107

SAFETY NOTES

 Even though you are working with low voltage, you are building safe habits which you will use later in high voltage circuits. Observe safety precautions when measuring voltage. Do not allow yourself to become part of the circuit, by touching the metal probes of the meter, or any part of the circuit.

PROCEDURES	FINDINGS	CONCLUSIONS
1. Place the ON/OFF switch on the front panel of the power supply to the OFF position.	—	—
2. Rotate the voltage adjust control on the front panel of the power supply fully CCW (counterclockwise). This will set the output of the power supply to zero volts.	—	—
3. Connect the circuit shown in Schematic 1. Place switches in the positions shown in Schematic 1. (S_1, S_4 closed, S_2, S_3 open)	—	—
4. Set up the meter to measure DC volts, and connect it across the output terminals of the power supply.	—	4. Observe the effect on the circuit, caused by the short circuit across R_1. Notice what happened to the voltage drop across R_1. The voltage drop across R_1 is now _____ volts, which indicates that the resistor is _____ (shorted/open). What happened to the voltage drop across R_2? _____ _____ _____ _____.
5. Reenergize the power supply, by placing the ON/OFF switch in the ON position. Adjust the voltage adjust control until the value measured on the voltmeter equals the value indicated on the schematic diagram.	—	—

Experiment 25

Troubleshooting Series Circuits Using Voltage Readings ■ 109

PROCEDURES	FINDINGS	CONCLUSIONS
6. Measure all values specified in Table 1, and record them in Table 1.	6. Record the readings obtained in the blanks below: V_{R1} —————— V_{R2} ——————	—
7. Place the ON/OFF switch on the front panel of the power supply to the OFF position.	—	—
8. Place a short across resistor R_1, by connecting an alligator test lead across the ends of the resistor.	8. Record the readings obtained in the blanks below V_{R1} —————— V_{R2} ——————	8. Observe the effect on the circuit, caused by the short circuit across R_2. Notice what happened to the voltage drop across R_2. The voltage drop across R_2 is now —————— volts, which indicates that the resistor is —————— (shorted/open). What happened to the voltage drop across R? —————————— —————————— —————————— ——————————.
9. Measure the voltage across R_1, and record the reading in Table 1. Observe the effect on the circuit, caused by the short circuit across R_1. Remove the alligator clip test lead.	9. Record the readings obtained in the blanks below: V_{R1} —————— V_{R2} ——————	—
10. Disconnect one end of resistor R_2, to create a simulated open resistor.	10. Record the readings obtained in the blanks below: V_{R1} —————— V_{R2} ——————	10. Observe the effect on the circuit, caused by R_1 being open. Notice what happened to the voltage drop across R_1. The voltage drop across R_1 is now —————— volts, which indicates that the resistor is —————— (shorted/open). (Note: This experiment simulates an open resistor. The resistor is not actually open, but S_1 gives the impression of the resistor being open. If the resistor

PROCEDURES	FINDINGS	CONCLUSIONS

were actually open, it would be obvious, since it would have to be physically damaged, or disconnected, to achieve this condition.) What happened to the voltage drop across R_2?

_____
_____
_____
_____

11. Measure the voltage across the circuit from point A to point B with the voltmeter. (Refer to Schematic 2.) Record the voltage reading in Table 1. Observe the effect on the circuit, caused by the open circuit between point A and point B.

11. Record the readings obtained in the blanks below:

V_{R1} _____

V_{R2} _____

11. Observe the effect on the circuit, caused by R_2 being open. Notice what happened to the voltage drop across R_2. The voltage drop across R_2 is now _____ volts, which indicates that the resistor is

_____

(shorted/open). (Note: This experiment simulates an open resistor. The resistor is not actually open, but S_4 gives the impression of the resistor being open. If the resistor were actually open, it would be obvious, since it would have to be physically damaged, or disconnected, to achieve this condition.) What happened to the voltage drop across R_1?

_____
_____
_____
_____

12. Obtain a test board from the instructor, which contains pre-selected failures (shorts across components, and open circuits between components). Practice troubleshooting these failures until you are proficient enough to take the troubleshooting test, if given by the instructor.

—

—

Experiment 26

TROUBLESHOOTING SERIES CIRCUITS USING CURRENT READINGS

SCHEMATIC:

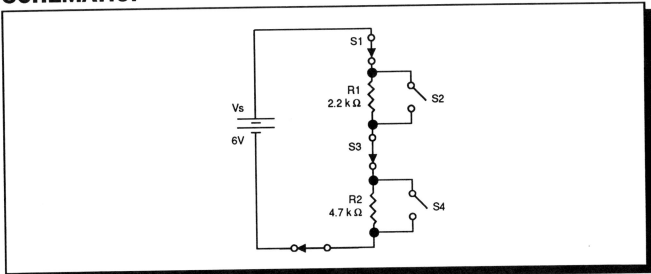

DISCUSSION

Although the technician does not directly measure current very often, it can be the simplest and most informative test for a particular situation. Because of the fact that the circuit must be broken to take a current measurement, it can be awkward to do in some situations. However, if there is a switch in the circuit, or a fuse holder, for example, to connect the ammeter across, then the current measurement becomes much easier. All that is necessary to take the measurement is to connect across the open switch contacts, or across the fuse terminals, with the fuse removed from the circuit.

TEXT CORRELATION

Before beginning this experiment, review section 9–6 on troubleshooting series circuits in *Electricity*.

OBJECTIVES

1. To troubleshoot a series circuit using current readings only, to determine what is malfunctioning in the circuit.

EQUIPMENT

- ☐ Standard tool box
- ☐ Standard parts kit
- ☐ Volt-ohm-milliammeter, 20 kilohm/volt DC OR
- ☐ Digital multimeter, with 10 megohm input Resistance on voltage scales
- ☐ Adjustable power supply, capable of 0 to 40 VDC output
- ☐ 5 single throw, single pole switches

PROCEDURES	FINDINGS	CONCLUSIONS
1. Place the ON/OFF switch on the front panel of the power supply in the OFF position.	—	—
2. Rotate the voltage adjust control on the front panel of the power supply fully CCW (counterclockwise). This will set the output of the power supply to zero volts.	—	—
3. Connect the circuit shown in Schematic 1. Place the switches in the positions shown on the schematic (S_1, S_3, and S_5 closed, S_2 and S_4 open.)	—	—
4. Reenergize the power supply, by placing the ON/OFF switch in the ON position. Adjust the voltage adjust control until the value measured on the voltmeter equals the value indicated on the schematic diagram.	—	—
5. Move switch S_5 to the open (off) position. Calculate the value of current that is normal for this circuit, and record the value.	5. Record the calculated value in the blank below: I_t _____	—
6. Connect the multimeter, set up as an ammeter, across the contacts of the S_5 switch. Measure and record the current reading, if any.	6. Record the reading obtained in the blank below: I_t _____	6. Notice what is happening in the circuit. Even though S_5 is open, which seemingly should deenergize the circuit, current is flowing through the ammeter. Can you explain why this is happening? _____ _____ _____ _____

Experiment 26   Troubleshooting Series Circuits Using Current Readings ■ 113

PROCEDURES	FINDINGS	CONCLUSIONS
7. Close S_5. Measure and record the current reading, if any.	7. Record the reading obtained in the blank below: I_t _____	7. Notice what the effect is on the current flowing in the circuit. Even though the circuit is energized, by closing S_5 there is no current being recorded on the ammeter. Why is this so? _____ _____ _____ _____.
8. With the ammeter still connected across the contacts of S_5, open S_5 and S_1. Measure and record the current reading, if any.	8. Record the reading obtained in the blank below: I_t _____	8. Notice the effect on current flowing in the circuit. The resistance in the circuit, because of the S_1 switch being open, has _____ (increased/decreased/decreased to zero) from the normal, which indicates a/an _____ (open circuit, short circuit, component out of tolerance).
9. Close S_1 and open S_3. Measure and record the current reading, if any.	9. Record the reading obtained in the blank below: I_t _____	9. Compare the results with the results you obtained in step 8. How do they compare? _____ _____ _____ _____.
10. Close S_3 and close S_2. Measure and record the current reading, if any.	10. Record the reading obtained in the blank below: I_t _____	10. Notice the effect on current flowing in the circuit. The resistance in the circuit, because of the S_2 switch being closed, has _____ (increased/decreased/decreased to zero) from the normal, which indicates a/an _____

PROCEDURES	FINDINGS	CONCLUSIONS
		(open circuit, short circuit, component out of tolerance).
11. Open S_2, and close S_4. Measure and record the current reading, if any.	11. Record the reading obtained in the blank below: I_t _____	11. Compare the results with the results you obtained in step 10. How do they compare? _____ _____ _____ _____

Experiment 27

SERIES CIRCUIT TROUBLESHOOTING PERFORMANCE TEST

SCHEMATIC:

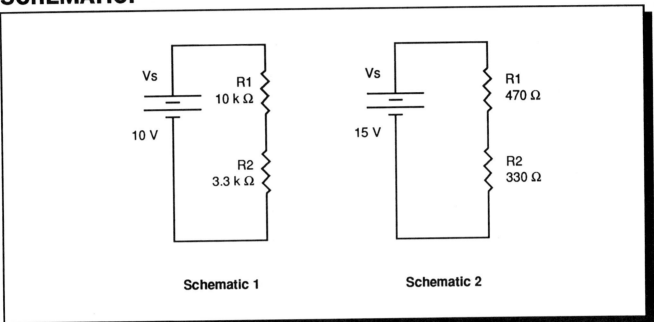

Schematic 1   Schematic 2

DISCUSSION

In this lab project, you will be given the opportunity to use the skills you have been developing over the last few lab projects. You will troubleshoot circuits using resistance readings and/or voltage readings. The troubleshooting procedures do not include current readings, because of the difficulty of connecting and disconnecting ammeters in the circuit.

TEXT CORRELATION

Before beginning this experiment, review section 9–6 on troubleshooting series circuits in *Electricity*.

OBJECTIVES

1. To test the student's ability to troubleshoot series circuits, using voltage and/or resistance readings.

EQUIPMENT

- ☐ Standard tool box
- ☐ Standard parts kit
- ☐ Adjustable power supply, capable of 0 to 40 VDC output
- ☐ Digital multimeter, with 10 megohm input resistance on voltage scales OR
- ☐ Volt-ohm-milliammeter, 20 kilohm/volt DC

PROCEDURES	FINDINGS	CONCLUSIONS
1. Obtain a test board from the instructor, which contains pre-selected failures (shorts across components, and open circuits between components).	—	—
2. Refer to the schematic diagram of the test board you received. Calculate all values of voltage, resistance, and/or current necessary for you to analyze the circuit. This is necessary, in order for you to know the correct values for this circuit. Then when you perform tests of the voltage or resistance, you will know which readings are normal and which are abnormal, and will be better able to interpret the readings you are getting.	—	—
3. When the instructor tells you to begin troubleshooting, perform whatever resistance tests and/or voltage tests you feel are necessary for the problem your circuit has. The test may be timed, at the discretion of the instructor.	3. Record the readings obtained in the blanks below: _____ _____ _____ _____	3. As you perform each test, think about what the reading is telling you. Ask yourself questions, like "If the voltage reading I just took is higher than the normal voltage reading at this point, what could be causing this?" or "What kind of circuit defect could cause the voltage across this resistor to equal the source voltage?" or maybe "What could be wrong with this circuit that would cause a voltage reading of 0 volts when I measure across this resistor?" or "Why does this resistance read so low across this resistor?"
4. When you have taken sufficient tests on the circuit to determine what the defect is (the circuit may even be functioning normally), notify the instructor that you have identified the malfunction, if any.	—	—

PROCEDURES	FINDINGS	CONCLUSIONS
5. Repeat this procedure for as many test boards as assigned by the instructor.	5. Record the readings obtained in the blanks below: _____ _____ _____ _____ Record the readings obtained in the blanks below: _____ _____ _____ _____	5. Remember to relate each voltage reading back to Ohm's Law, which states that voltage readings are the product of two quantities — resistance and current. As you measure each voltage, realize that if it is high, it is high because of either too much current or too much resistance. If the voltage is too low, it is too low because of either too little current or too little resistance.

Experiment 28

MAXIMUM TRANSFER OF POWER

SCHEMATIC:

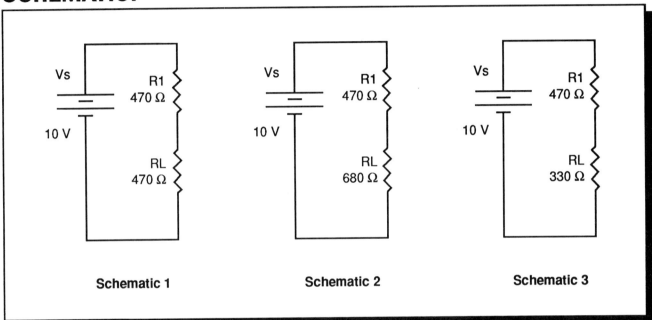

Schematic 1   Schematic 2   Schematic 3

DISCUSSION

In nearly all amplifier circuits, the design is based on one of two goals: to transfer maximum voltage to the next amplifier stage, or to transfer maximum power to the next stage or to the load. In this project, we will look at what is necessary to achieve maximum transfer of power, by studying the relationship of internal resistance of a voltage source to the resistance of the load connected to that voltage source.

TEXT CORRELATION

Before beginning this experiment, review section 7–4 on calculating power in *Electricity*.

OBJECTIVES

1. To build and test a circuit to demonstrate the concept of maximum power transfer.

2. To observe the effect of having a load resistance that is not matched to the source resistance.

EQUIPMENT

- ☐ Standard tool box
- ☐ Standard parts kit
- ☐ Adjustable power supply, capable of 0 to 40 VDC
- ☐ Digital multimeter, with 10 megohm input resistance on voltage scales OR
- ☐ Volt-ohm-milliammeter, 20 kilohm/volt DC

120 ■ Maximum Transfer of Power

Experiment 28

PROCEDURES	FINDINGS	CONCLUSIONS
1. Place the ON/OFF switch on the front panel of the power supply to the OFF position.	—	—
2. Rotate the voltage adjust control on the front panel of the power supply fully CCW (counterclockwise). This will set the output of the power supply to zero volts.	—	—
3. Connect the circuit shown in Schematic 1.	—	—
4. Calculate total resistance, total current, and the voltage drops across the load resistor, and the internal resistance of the power supply (represented by the R_1 resistor).	4. Record the calculated values in the blanks below: R_t _____ I_t _____ V_{R1} _____ V_{RL} _____	—
5. With the circuit connected, reenergize the power supply, and adjust the supply until the output matches the value indicated on the schematic, as measured by a multimeter connected as a voltmeter across the terminals of the power supply.	—	—
6. Measure and record the values for I_t, V_{R1} and V_{RL} in the blanks indicated.	6. Record the readings obtained in the blanks below I_t _____ V_{R1} _____ V_{RL} _____	—
7. Using the formula P=IV, calculate the amount of power that will be dissipated by the load resistor (R_L)	7. Record the calculated value in the blank below: Load Power _____	—
8. Replace the load resistor with the new value of load resistor indicated in Schematic 2.	—	—

Experiment 28 Maximum Transfer of Power ■ 121

PROCEDURES	FINDINGS	CONCLUSIONS
9. Repeat steps 4 through 7 for the new value of load resistor.	9. Record the calculated values in the blanks below: R_t _____ I_t _____ V_{R1} _____ V_{RL} _____ Record the readings obtained in the blanks below: I_t _____ V_{R1} _____ V_{RL} _____ Record the calculated value in the blank below: Load Power _____	9. Compare the amount of power dissipated by the load resistor in step 7 with the amount of power dissipated in step 9. _____ _____ _____ _____.
10. Replace the load resistor with the new value of load resistor indicated in Schematic 3.		
11. Repeat steps 4 through 7 for the new value of load resistor.	11. Record the calculated values in the blanks below: R_t _____ I_t _____ V_{R1} _____ V_{RL} _____ Record the readings obtained in the blanks below: I_t _____ V_{R1} _____ V_{RL} _____ Record the calculated value in the blank below: Load Power _____	11. Compare the amount of power dissipated by the load resistor in step 7 with the amount of power dissipated in step 11 and step 9. _____ _____ _____ _____. What conclusion does this data suggest to you? _____ _____ _____ _____.

Experiment 29

ANALYZING PARALLEL CIRCUITS

SCHEMATIC:

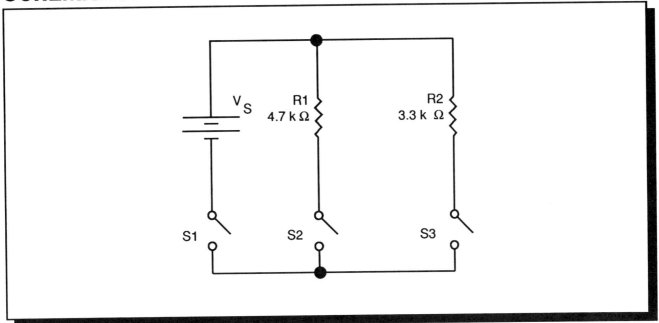

DISCUSSION

The parallel connection is the second way of hooking up components in electronic circuits. Most of the rules we studied for series circuits will not work in parallel circuits. For example, in series circuits we can simply add up the values of the resistors to get total resistance. In parallel, as we add resistors to the circuit, the resistance actually goes down. In series circuits, the current is the same everywhere, but in parallel circuits, the current is different in each branch. In series circuits, the voltage may be different across each resistor in the circuit, while in parallel circuits the voltage is exactly the same across each of the components.

TEXT CORRELATION

Before beginning this experiment, review section 10-3 on Kirchoff's Current Law in *Electricity*.

OBJECTIVES

1. To connect a parallel circuit, and determine the values of current, voltage and resistance.

EQUIPMENT

- ☐ Standard tool box
- ☐ Standard parts kit
- ☐ Adjustable power supply, capable of 0 to 40 VDC output
- ☐ Digital multimeter, with 10 megohm input resistance on voltage scales OR
- ☐ Volt-ohm-milliammeter, 20 kilohm/volt DC

123

SAFETY NOTES

 Remember that the Ohmmeter should never be used in a circuit that has power applied to it. Always turn off the power supply and disconnect the leads from the circuit before measuring the resistance in the circuit.

PROCEDURES	FINDINGS	CONCLUSIONS
1. Place the ON/OFF switch on the front panel of the power supply to the OFF position.	—	—
2. Rotate the voltage adjust control on the front panel of the power supply fully CCW (counterclockwise). This will set the output of the power supply to zero volts.	—	—
3. Connect the circuit shown in Schematic 1.	—	—
4. Calculate the values of R_t, I_1, I_2, and I_t. Record the values in the blanks indicated.	4. Record the calculated values in the blanks below: R_t _____ I_1 _____ I_2 _____ I_t _____	4. Notice that the total current is equal to the _____ (sum/difference) of the individual branch currents. Also, the total resistance of the circuit is _____ (less/more) than the _____ (smallest/largest) resistor.
5. Reenergize the power supply, by placing the ON/OFF switch in the ON position. Adjust the voltage adjust control until the value measured on the voltmeter equals the value indicated on the schematic diagram. Measure and record the values for R_t, I_1, I_2, and I_t in the blanks indicated. In order to measure I_1, I_2, and I_t, simply connect the ammeter across S_1, S_2, or S_3 as appropriate, then open the applicable switch to allow current to flow through the ammeter.	5. Record the readings obtained in the blanks below: R_t _____ I_1 _____ I_2 _____ I_t _____	5. Did your measured values equal the values you calculated? _____ If there are differences, what might be causing them? _____ _____ _____

Experiment 30

TROUBLESHOOTING PARALLEL CIRCUITS

SCHEMATIC:

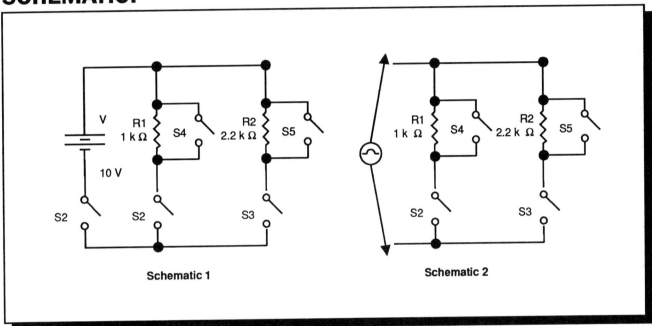

Schematic 1

Schematic 2

DISCUSSION

Troubleshooting parallel circuits is more difficult than troubleshooting series circuits, partly because it is useless in most cases to try to troubleshoot using voltage readings. This is true simply because the voltage across all components in parallel is the same, whether the component is good or bad. For example, if you measure the voltage across a resistor that is open in a parallel circuit, you will still measure the exact same voltage as across all of the other components in parallel. If a component connected in parallel should happen to short, it will cause the technician to measure 0 ohms across each of the components in the circuit — good or bad.

TEXT CORRELATION

Before beginning this experiment, review section 10–4 on troubleshooting parallel circuits in *Electricity*.

OBJECTIVES

1. To troubleshoot a parallel circuit, to determine the malfunction.

EQUIPMENT

- ☐ Standard tool box
- ☐ Standard parts kit
- ☐ Adjustable power supply, capable of 0 to 40 VDC
- ☐ Digital multimeter, with 10 megohm input resistance on voltage scales OR
- ☐ Volt-ohm-milliammeter, 20 kilohm/volt DC

PROCEDURES	FINDINGS	CONCLUSIONS
1. Place the ON/OFF switch on the front panel of the power supply to the OFF position.	—	—
2. Rotate the voltage adjust control on the front panel of the power supply fully CCW (counterclockwise). This will set the output of the power supply to zero volts.	—	—
3. Connect the circuit shown in Schematic 1. Place switches S_1, S_2, and S_3 in the closed position; switches S_4 and S_5 must remain open. Calculate and record the value of current that should flow through the circuit.	3. Record the calculated value in the blank below: I_t _____	—
4. Reenergize the power supply, by placing the ON/OFF switch in the ON position. Adjust the voltage adjust control until the value measured on the voltmeter equals the value indicated on the schematic diagram.	—	—
5. Connect a multimeter set up as an ammeter across the contacts of switch S_1. Open switch S_1, then observe and record the reading on the ammeter.	5. Record the reading obtained in the blank below: I_t _____	5. Compare the reading obtained to the calculated value. Are they the same? _____ If not, what could account for the differences in the values? _____ _____ _____ _____.
6. Open switch S_2, simulating an open in the R_1 branch. Measure and record the value for It in the blanks indicated.	6. Record the reading obtained in the blank below: I_t _____	6. Compare the new value of current to that measured in step 5. Now the current has _____ (increased/decreased) to _____ A, because of the open in the R_1 branch.

Experiment 30 — Troubleshooting Parallel Circuits

PROCEDURES	FINDINGS	CONCLUSIONS
7. Close switch S_2, restoring the circuit, and then open switch S_3, simulating an open in the R_2 branch. Measure and record the value for It in the blanks indicated.	7. Record the reading obtained in the blank below: I_t _____	7. Compare the new value of current to that obtained in step 6. Is the new value more or less? _____ Why? _____ _____ _____ _____.
8. Close switch S_3 to restore the circuit to normal. Place the ON/OFF switch on the front panel of the power supply to the OFF position.	—	—
9. Disconnect the power supply from the circuit. Calculate and record the normal value of resistance in the circuit. Connect the multimeter, set up as an ohmmeter, across the terminals where the power supply had been connected. Record the reading on the ohmmeter in the indicated blank.	9. Record the calculated value in the blank below: R_t _____ Record the reading obtained in the blank below: R_t _____	
10. Open switch S_2, and observe and record the resistance.	10. Record the reading obtained in the blank below: R_t _____	10. How was the resistance affected when S_2 was opened? _____ Why? _____ _____ _____ _____.
11. Close S_2 and open S_3. Measure and record the value for R_t in the blank indicated.	11. Record the reading obtained in the blank below: R_t _____	11. Compare this new value of resistance to the reading obtained in step 9? _____ _____ _____ _____ _____.

PROCEDURES

12. Close S_3 and S_4, simulating a shorted resistor R_1. Measure and record the value for R_t in the blank indicated.

13. Open S_4 and close S_5. Measure and record the value for R_t in the blank indicated.

FINDINGS

12. Record the reading obtained in the blank below:

    R_t _____

13. Record the reading obtained in the blank below:

    R_t _____

CONCLUSIONS

12. Now what happened to the resistance? _____
    Why would this happen?
    _____
    _____
    _____
    _____

13. Compare this reading to step 12.
    _____
    _____
    _____
    _____

Experiment 31

DESIGNING PARALLEL CIRCUITS

SCHEMATIC:

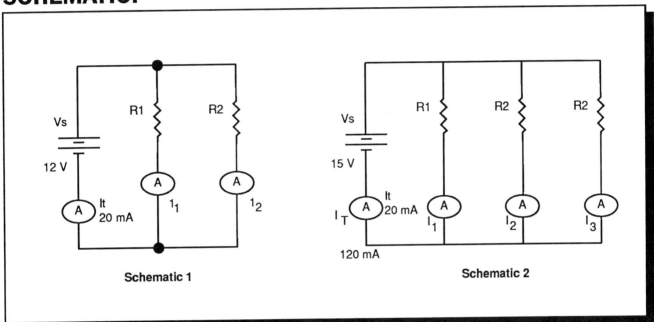

DISCUSSION

Parallel circuits are used in many applications. One of the most common is in house wiring. All of the wall sockets in the home are wired in parallel. The reason is that parallel branches are relatively independent of each other. Plugging in the TV doesn't normally affect the operation of the refrigerator (so long as the TV set doesn't overload the circuit and cause the circuit breaker to trip). Also, since all the circuitry is in parallel, all of the circuits operate with the same voltage. That makes the design of the appliances simpler, since the only voltage necessary to design for is 115 volts.

TEXT CORRELATION

Before beginning this experiment, review section 10–4 on troubleshooting parallel circuits in *Electricity*.

OBJECTIVES

1. To design a parallel circuit to a set of specifications.
2. To build the circuit and test it.

EQUIPMENT

- ☐ Standard tool box
- ☐ Standard parts kit
- ☐ Adjustable power supply, capable of 0 to 40 VDC
- ☐ Digital multimeter, with 10 megohm input Resistance on voltage scales OR
- ☐ Volt-ohm-milliammeter, 20K ohm/volt DC

130 ■ Designing Parallel Circuits

Experiment 31

| **PROCEDURES** | **FINDINGS** | **CONCLUSIONS** |

Design Problem 1

—   —

1. Using standard sizes from the Standard Parts Kit, design a parallel circuit to have the following specifications. (Refer to Schematic 1.) ratio of current in branch 1 to current in branch 2 = 2:1, I_{total} = 20 mA V_{in} = 12 volts

2. Record the values of the resistors necessary to build this circuit, including wattage rating.

2. Record the calculated values in the blanks below:

R_1 _____@

_____watts

R_2 _____@

_____watts

—

3. Calculate the current in each branch and total current. Enter the values in the blanks.

3. Record the calculated values in the blanks below:

I_1 _____

I_2 _____

I_t _____

—

4. Connect the circuit, using standard sizes from the Standard Parts Kit. Measure the branch currents and the total current, and enter in the blanks.

4. Record the readings obtained in the blanks below:

I_1 _____

I_2 _____

I_t _____

4. How close were the measured values to the calculated values?

_____

_____

_____

_____.

How do you account for the differences in the actual values compared to the calculated values?

_____

_____

_____

_____.

Experiment 31 — Designing Parallel Circuits ■ 131

PROCEDURES	FINDINGS	CONCLUSIONS

Design Problem 2

5. Using standard sizes from the Standard Parts Kit, design a parallel circuit to have the following specifications. (Refer to Schematic 2.) ratio of current $I_1:I_2:I_3$ equals 1:2:3

   I_{total} = 120 mA

   V_{in} = 15 volts

6. Record the values of the resistors necessary to build this circuit, including wattage rating.

6. Record the calculated values in the blanks below:

   R_1 _____ @

   _____ watts

   R_2 _____ @

   _____ watts

   R_3 _____ @

   _____ watts

7. Calculate the current in each branch and total current. Enter the values in the blanks.

7. Record the calculated values in the blanks below:

   I_1 _____

   I_2 _____

   I_3 _____

   I_t _____

8. Connect the circuit, using standard sizes from the Standard Parts Kit. Measure the branch currents, and the total current, and enter in the blanks.

8. Record the readings obtained in the blanks below:

   I_1 _____

   I_2 _____

   I_3 _____

   I_t _____

8. How close were the measured values to the calculated values?

_____

_____

_____

_____.

How do you account for the differences in the actual values compared to the calculated values?

_____

_____

_____

_____.

Experiment 32

ANALYZING SERIES-PARALLEL COMBINATION CIRCUITS

SCHEMATIC:

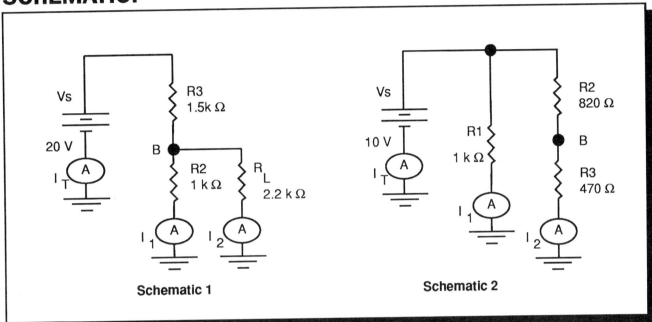

Schematic 1

Schematic 2

DISCUSSION

Most electronic circuits are constructed such that they are actually a combination series-parallel circuit. That is, parts of the circuit are connected in series, while other portions of the circuit are connected in parallel. The rules you studied in the section of the course covering series circuits will be used for the portion of the circuit that is series connected. The rules for parallel circuits will be used for the portion of the circuit that is parallel connected.

TEXT CORRELATION

Before beginning this experiment, review section 11–2 on simplifying reconstruct method in *Electricity.*

OBJECTIVES

1. To build and test a series-parallel combination circuit, to determine the operating characteristics of the circuit.

EQUIPMENT

- [ ] Standard tool box
- [ ] Standard parts kit
- [ ] Adjustable power supply, capable of 0 to 40 VDC
- [ ] Digital multimeter, with 10 megohm input resistance on voltage scales OR
- [ ] Volt-ohm-milliammeter, 20 kilohm/volt DC

133

134 ■ Analyzing Series-Parallel Combination Circuits

PROCEDURES	FINDINGS	CONCLUSIONS

1. Place the ON/OFF switch on the front panel of the power supply in the OFF position.

2. Rotate the voltage adjust control on the front panel of the power supply fully CCW (counterclockwise). This will set the output of the power supply to zero volts.

3. Connect the circuit shown in Schematic 1.

4. Calculate the values of resistance, current, and voltage in the circuit. Record the indicated values in the blanks.

   4. Record the calculated values in the blanks below:

   R_t ——————

   I_t ——————

   I_1 ——————

   I_2 ——————

   V_{RL} ——————

   V_{R2} ——————

   V_{R3} ——————

   V_B ——————

5. Reenergize the power supply, by placing the ON/OFF switch in the ON position. Adjust the voltage adjust control until the value measured on the voltmeter equals the value indicated on the schematic diagram.

6. Measure and record the values for I_t, I_1, I_2, V_{R1}, V_{R2}, V_{R3}, and V_B in the blanks indicated.

   6. Record the readings obtained in the blanks below:

   I_t ——————

   I_1 ——————

   I_2 ——————

   V_{R1} ——————

   V_{R2} ——————

   V_{R3} ——————

   V_B ——————

   6. Compare the readings obtained to the values calculated. How do they compare?

   How do you account for the differences?

   ——————————

   ——————————

   ——————————

   ——————————

PROCEDURES	FINDINGS	CONCLUSIONS
7. Place the ON/OFF switch on the front panel of the power supply in the OFF position.	—	—
8. Rotate the voltage adjust control on the front panel of the power supply fully CCW (counterclockwise). This will set the output of the power supply to zero volts.	—	—
9. Connect the circuit shown in Schematic 2.	—	—
10. Calculate the values of resistance, current, and voltage in the circuit. Record the indicated values in the blanks.	10. Record the calculated values in the blanks below: R_t ——— I_t ——— I_1 ——— I_2 ——— V_{R1} ——— V_{R2} ——— V_{R3} ——— V_B ———	—
11. Reenergize the power supply, by placing the ON/OFF switch in the ON position. Adjust the voltage adjust control until the value measured on the voltmeter equals the value indicated on the schematic diagram.	—	—
12. Measure and record the values for I_t, I_1, I_2, V_{R1}, V_{R2}, V_{R3}, and V_B in the blanks indicated.	12. Record the readings obtained in the blanks below: I_t ——— I_1 ——— I_2 ——— V_{R1} ——— V_{R2} ——— V_{R3} ——— V_B ———	12. Compare the readings obtained to the values calculated. How do they compare? ——— ——— ——— ——— How do you account for the differences? ——— ———

Experiment 33

LOADED VOLTAGE DIVIDERS

SCHEMATIC:

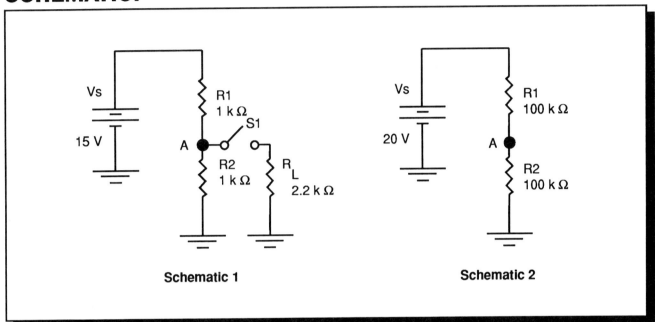

Schematic 1

Schematic 2

DISCUSSION

Voltage dividers are used regularly in electronics circuits, for providing several different voltages from the one fixed power supply voltage. This is necessary since different values of voltage are needed to operate various parts of the equipment. For example, it is generally true that the higher the DC voltage applied to a transistor amplifier, the more noise produced by the circuit. Therefore, it is practical to provide the input stages (when the signal to be processed is small) with a lower voltage.

TEXT CORRELATION

Before beginning this experiment, review section 11–1 on simplifying series parallel circuits in *Electricity*.

OBJECTIVES

1. To build a voltage divider circuit, with a load connected to it.

2. To analyze the effect of loading the circuit.

EQUIPMENT

- ☐ Standard tool box
- ☐ Standard parts kit
- ☐ Adjustable power supply, capable of 0 to 40 VDC
- ☐ Digital multimeter, with 10 megohm input resistance on voltage scales AND
- ☐ Volt-ohm-milliammeter, 20 kilohm/volt DC

138 ■ Loaded Voltage Dividers Experiment 33

PROCEDURES	FINDINGS	CONCLUSIONS
1. Place the ON/OFF switch on the front panel of the power supply to the OFF position.	—	—
2. Rotate the voltage adjust control on the front panel of the power supply fully CCW (counterclockwise). This will set the output of the power supply to zero volts.	—	—
3. Calculate the value of voltage that should be measured at point A in the circuit of Schematic 1, with switch S_1 open. (This simulates an unloaded voltage divider circuit.)	3. Record the calculated value in the blank below: _____	—
4. Measure and record the value for voltage from A to ground in the blank indicated.	4. Record the reading obtained in the blank below: _____	4. How does the calculated value compare to the measured value? _____ How do you account for the difference in the values? _____ _____ _____ _____.
5. Close S_1. Measure and record the value for voltage at point A in respect to ground, in the blanks indicated. (This will simulate the concept known as loading.)	5. Record the reading obtained in the blank below: _____	5. What effect did you observe when you closed the switch? _____ _____ _____ _____. Can you explain why this happened? _____ _____ _____ _____.

Experiment 33   Loaded Voltage Dividers ■ 139

PROCEDURES	FINDINGS	CONCLUSIONS

6. Place the ON/OFF switch on the front panel of the power supply to the OFF position.

  —   —

7. Rotate the voltage adjust control on the front panel of the power supply fully CCW (counterclockwise). This will set the output of the power supply to zero volts.

  —   —

8. Connect the circuit shown in Schematic 2.

  —   —

9. Calculate and record the expected voltage reading from point A to ground.

  9. Record the calculated value in the blank below:

  _____

  —

10. Reenergize the power supply, by placing the ON/OFF switch in the ON position. Adjust the voltage adjust control until the value measured on the voltmeter equals the value indicated on the schematic diagram.

  —   —

11. Using the DMM, measure and record the value for voltage from point A to ground in the blank indicated.

  11. Record the reading obtained in the blank below:

  _____

  11. How does the calculated value compare to the measured value?

  _____

  How do you account for the difference in the values?

  _____
  _____
  _____
  _____.

12. Using the VOM, set on the lowest scale that will allow measurement, measure and record the voltage from point A to ground.

  12. Record the reading obtained in the blank below:

  _____

  12. How does the calculated value compare to the measured value?

  _____

  How do you account for the difference in the values?

  _____
  _____
  _____
  _____.

PROCEDURES	FINDINGS	CONCLUSIONS
		This effect is referred to as meter "loading effect". It can be reduced by moving to the next higher range, since this increases the internal resistance of the meter.

Experiment 34

TROUBLESHOOTING SERIES-PARALLEL CIRCUITS USING RESISTANCE MEASUREMENTS

SCHEMATIC:

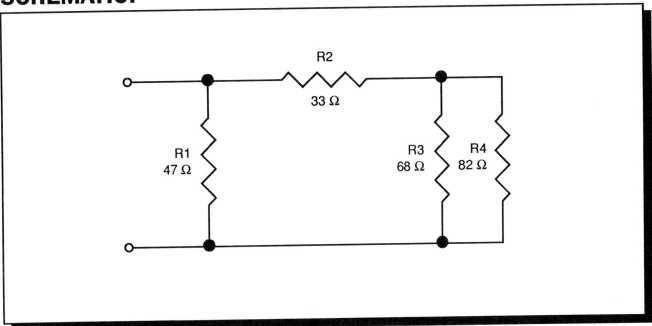

DISCUSSION

If the technician can troubleshoot series-parallel combination circuits using resistance readings, he/she will be able to repair complicated electronics equipment with little difficulty. Many technicians simply use signal injection or signal tracing techniques, which you will learn about later, until they determine the defective stage (each stage being a transistor, vacuum tube, or integrated circuit and its associated components), then turn off the power and begin measuring resistances in the circuit to determine the defective component.

TEXT CORRELATION

Before beginning this experiment, review section 11–4 on troubleshooting series parallel circuits in *Electricity*.

OBJECTIVES

1. To troubleshoot series-parallel combination circuits, using resistance readings only, and determine which is the defective component.

EQUIPMENT

- ☐ Standard tool box
- ☐ Standard parts kit
- ☐ Digital multimeter, with 10 megohm input resistance on voltage scales AND
- ☐ Volt-ohm-milliammeter, 20 kilohm/volt DC

PROCEDURES	FINDINGS	CONCLUSIONS
1. Obtain a test board from the instructor that contains pre-selected failures (shorts across components, and open circuits between components).	—	—
2. Troubleshoot the test board, using resistance readings only, to determine what the defect is, if any. (The PC board may not have a defect in it.)	2. Record test results in the troubleshooting table.	—
3. When you have determined, using resistance readings, what the defect is, notify the instructor.	—	3. The defect is: _____ _____ _____ _____
4. Repeat this procedure for any additional troubleshooting tests the instructor may require. (Suggested number of troubleshooting tests is three.)	—	—

Test Type_____
Steps taken to find defect:
_____
_____
_____
_____
_____
_____

Test Type_____
Steps taken to find defect:
_____
_____
_____
_____
_____
_____

Test Type_____
Steps taken to find defect:
_____
_____
_____
_____
_____
_____

Test Type_____
Steps taken to find defect:
_____
_____
_____
_____
_____
_____

Experiment 35

TROUBLESHOOTING SERIES-PARALLEL CIRCUITS USING VOLTAGE MEASUREMENTS

SCHEMATIC:

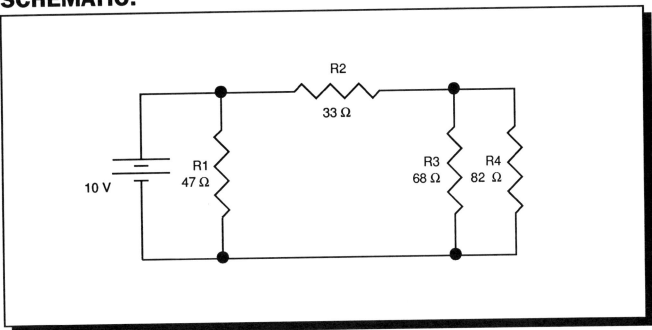

DISCUSSION

The technician must be able to troubleshoot series-parallel combination circuits using voltage readings. When the technician has determined which stage contains the defect, he/she may then begin measuring voltages around the circuit to determine where the defect is.

TEXT CORRELATION

Before beginning this experiment, review section 11–4 on troubleshooting series parallel circuits in *Electricity*.

OBJECTIVES

1. To troubleshoot series-parallel combination circuits, using voltage readings only, and determine which is the defective component.

EQUIPMENT

- ☐ Standard tool box
- ☐ Standard parts kit
- ☐ Adjustable power supply, capable of 0 to 40 VDC
- ☐ Digital multimeter, with 10 megohm input resistance on voltage scales OR
- ☐ Volt-ohm-milliammeter, 20 kilohm/volt DC

PROCEDURES | FINDINGS | CONCLUSIONS

1. Obtain a test board from the instructor, which contains pre-selected failures (shorts across components, and open circuits between components).

2. Troubleshoot the test board, using voltage readings only, to determine what the defect is, if any. (The PC board may not have a defect in it.)

    2. Record test results in the troubleshooting table.

3. When you have determined, using voltage readings, what the defect is, notify the instructor.

    3. The defect is:
    _____
    _____
    _____
    _____

4. Repeat this procedure for any additional troubleshooting tests the instructor may require. (Suggested number of troubleshooting tests is three.)

Test Type_____
Steps taken to find defect:
_____
_____
_____
_____
_____
_____

Test Type_____
Steps taken to find defect:
_____
_____
_____
_____
_____
_____

Test Type_____
Steps taken to find defect:
_____
_____
_____
_____
_____
_____

Test Type_____
Steps taken to find defect:
_____
_____
_____
_____
_____
_____

Experiment 36

TROUBLESHOOTING SERIES-PARALLEL CIRCUITS USING CURRENT MEASUREMENTS

SCHEMATIC:

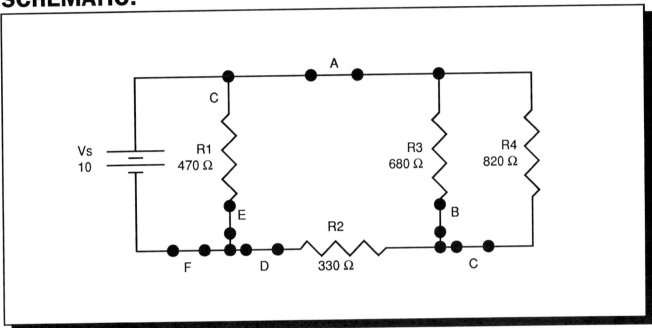

DISCUSSION

Although the technician does not use current readings as often as voltage or resistance tests for troubleshooting, they are a vital part of the technicians tools for troubleshooting. Quite often, the current-reading test is the easiest and most informative test to perform.

TEXT CORRELATION

Before beginning this experiment, review section 11–4 on troubleshooting series parallel circuits in *Electricity*.

OBJECTIVES

1. To troubleshoot series-parallel combination circuits, using current readings only, and determine what the defective component is.

EQUIPMENT

- ☐ Standard tool box
- ☐ Standard parts kit
- ☐ Adjustable power supply, capable of 0 to 40 VDC
- ☐ Digital multimeter, with 10 megohm input resistance on voltage scales AND
- ☐ Volt-ohm-milliammeter, 20 kilohm/volt DC

145

PROCEDURES	FINDINGS	CONCLUSIONS
1. Obtain a test board from the instructor that contains pre-selected failures (shorts across components, and open circuits between components).	—	—
2. Troubleshoot the test board, using current readings only, to determine what the defect is, if any. (The PC board may not have a defect in it.)	2. Record test results in the troubleshooting table.	—
3. When you have determined, using current readings, what the defect is, notify the instructor.	—	3. The defect is: _____ _____ _____ _____
4. Repeat this procedure for any additional troubleshooting tests the instructor may require. (Suggested number of troubleshooting tests is three.)	—	—

Test Type_____
Steps taken to find defect:
_____
_____
_____
_____
_____

Test Type_____
Steps taken to find defect:
_____
_____
_____
_____
_____

Test Type_____
Steps taken to find defect:
_____
_____
_____
_____
_____

Test Type_____
Steps taken to find defect:
_____
_____
_____
_____
_____

Experiment 37

TROUBLESHOOTING PERFORMANCE TEST FOR SERIES-PARALLEL CIRCUITS

SCHEMATIC:

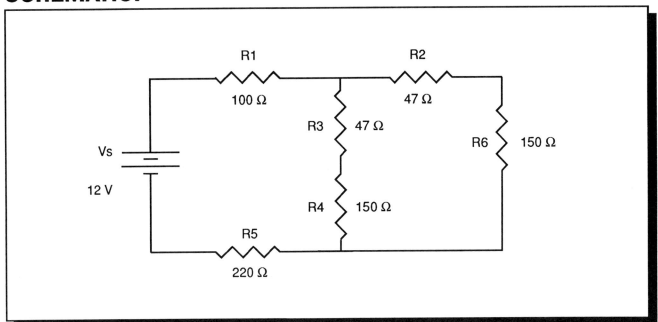

DISCUSSION

The single most important job the electronics technician performs is to troubleshoot circuitry to determine what defects might be present. Technicians at all levels, from entry level technicians to advanced technicians with many years' experience, must be able to troubleshoot. The two most common methods of troubleshooting down to the defective component are voltage and resistance readings. These two methods are emphasized in this lab project.

TEXT CORRELATION

Before beginning this experiment, review section 11-4 on troubleshooting series parallel circuits in *Electricity*.

OBJECTIVES

1. To troubleshoot series-parallel combination circuits, using resistance readings and/or voltage readings to determine what the defective component is.

EQUIPMENT

- ☐ Standard tool box
- ☐ Standard parts kit
- ☐ Adjustable power supply, capable of 0 to 40 VDC
- ☐ Digital multimeter, with 10 megohm input resistance on voltage scales AND
- ☐ Volt-ohm-milliammeter, 20 kilohm/volt DC

PROCEDURES | FINDINGS | CONCLUSIONS

1. Obtain a test board from the instructor that contains pre-selected failures (shorts across components, and open circuits between components). — —

2. Troubleshoot the test board, using resistance readings and/or voltage readings only, to determine what the defect is, if any. (The PC board may not have a defect in it.)

   2. Record test results in the troubleshooting table. —

3. When you have determined, using resistance readings and/or voltage readings, what the defect is, notify the instructor.

   —

   3. The defect is:
   _____
   _____
   _____
   _____

4. Repeat this procedure for any additional troubleshooting tests the instructor may require. (Suggested number of troubleshooting tests is three.) — —

Test Type_____
Steps taken to find defect:
_____
_____
_____
_____
_____
_____

Test Type_____
Steps taken to find defect:
_____
_____
_____
_____
_____
_____

Test Type_____
Steps taken to find defect:
_____
_____
_____
_____
_____
_____

Test Type_____
Steps taken to find defect:
_____
_____
_____
_____
_____
_____

Experiment 38

CAPACITOR COLOR CODE AND RC TIME CONSTANTS

SCHEMATIC:

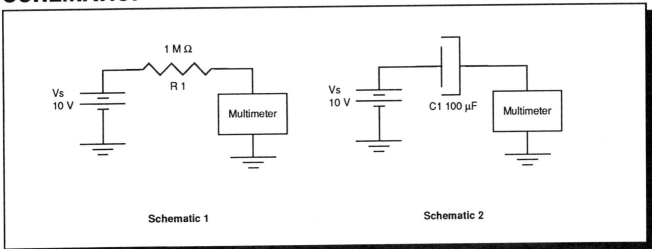

Schematic 1   Schematic 2

DISCUSSION

A capacitor is a planned "open circuit". As you already know, current cannot flow through an open circuit. However, because of the fact that the capacitor is constructed with conductive plates, there is a surplus of electrons available to allow current flow in the external circuit and charge the capacitor. This takes a period of time, determined by the value of resistor and the capacitor in the circuit. After a period of time, five time constants, the capacitor will be totally charged and there will be no move ment of electrons in the circuit.

TEXT CORRELATION

Before beginning this experiment, review section 15–3 on RC time constants in *Electricity*.

OBJECTIVES

1. To identify capacitors through the capacitor identification codes.

2. To measure the voltage in an RC circuit while charging and discharging a capacitor.

3. To observe the effect on the total capacitance of connecting capacitors in series and in parallel.

EQUIPMENT

- ☐ Standard tool box
- ☐ Adjustable power supply, capable of 0 to 40 VDC output
- ☐ Digital multimeter, with 10 megohm input resistance on voltage scales OR
- ☐ Volt-ohm-milliammeter, 20 kilohm/volt DC
- ☐ Assorted sizes and types of capacitors to be used in the capacitor identification portion of the experiment: (2) 100 µF electrolytic capacitors, 25 WVDC or higher (1) 1 megohm resistor, 1/4 watt or higher (1) alligator clip test lead.

SAFETY NOTES

Even though the voltage applied to the circuit is low, treat the capacitor with respect! A charged capacitor can shock you. Always discharge a capacitor before handling it or before troubleshooting the circuit it is installed in.

PROCEDURES	FINDINGS	CONCLUSIONS

1. Refer to the charts supplied and determine which of the capacitor codes apply to the capacitors you have been assigned to use in this experiment.

   —

   1. Use capacitors that are of several types — paper, tubular, mylar, polystyrene, mica, and ceramic capacitors. See your instructor if you do not know which types to use.

2. Using the appropriate capacitor identification code, identify the capacitors you have been given.

   2. List the values of the capacitors in the blanks below:
   1. _____
   2. _____
   3. _____
   4. _____

   2. At this point you are primarily interested in the value of capacitance, tolerance, and working voltage. The appropriate capacitance codes are listed in the text book in the section on capacitors.

3. Set up your multimeter on the appropriate scale to measure a minimum of 10 volts DC. Connect the circuit shown in Schematic 1, with the multimeter connected as shown. Energize the voltage source, and observe the DC voltage on the multimeter.

   3. Record the voltage measured on the multimeter in the blank below:

   _____

   3. Expect to see approximately 9.09 volts if you are using a digital multimeter, or approximately 1.67 volts if you are using an analog meter, on the 10 volt scale, if the sensitivity of the meter is 20 kilohms per volt.

4. Applying the following formula, determine the internal resistance of the meter you are using.

   4. $R_m = V_m R1/(V_s - V_m)$ where R_m = meter resistance, V_m = Voltage reading on the meter, and V_s = power supply voltage.
   R_m _____

   4. The internal resistance of most digital multimeters is 10 megohms, and 200 kilohms for an analog meter.

5. Connect the 100 µF capacitor in the circuit in place of R_1. Leave meter set on DC volts as before. (Refer to Schematic 2.)

   —

   —

6. Using an alligator test lead from the Standard Tool Box, connect across the leads of the capacitor to discharge it.

   —

   —

Experiment 38 Capacitor Code and RC Time Constants ■ 151

| **PROCEDURES** | **FINDINGS** | **CONCLUSIONS** |

7. Remove the short from across the capacitor, and observe the second hand on the clock, or the time on a stop watch, if available. Measure the time it takes for the voltage reading on the meter to decrease to 0 volts.

7. Time _____
 (1st attempt)

7. It is difficult to get an exact measurement of the voltage, because it is changing all the time. It is not necessary to obtain extreme accuracy, since the tolerance of the capacitor is usually +80% and -20% anyway. It is the concept that is important. How would the size of the resistor in series affect the reading you obtain?

_____
_____
_____
_____.

8. Once again, short across the leads of the capacitor. Repeat step 7 a second time, and record the time required to charge the capacitor.

8. Time _____
 (2nd attempt)

—

9. Repeat step 7 for the third time, and record the time required to charge the capacitor.

9. Time _____
 (3rd attempt)

—

10. Find the average of the three readings, to determine the average time to charge the capacitor.

10. Average time = (1st time + 2nd time + 3rd time)/3=

_____
_____.

10. If you are using a digital meter, the internal resistance of the meter is probably 10 megohms. Therefore, the time recorded should be approximately 500 seconds (8 minutes, 20 seconds). How will the time be affected if you are using a VOM with 20 kilohms per volt on the 10 volt range?

_____
_____
_____
_____.

11. Connect the second 100 µF capacitor across the first capacitor (parallel connection).

—

—

PROCEDURES	FINDINGS	CONCLUSIONS
12. Repeat steps 6 and 7.	12. Record the amount of time necessary to charge the capacitor. _____	12. Since there are now 2 equal size capacitors in parallel, the total capacitance has _____ _____ (increased, decreased) to _____ µF.
13. Disconnect the second 100 µF capacitor and reconnect it in series with the first 100 µF capacitor.	—	—
14. Repeat steps 6 and 7.	14. Record the amount of time necessary to charge the capacitor. _____	14. Since there are now 2 equal size capacitors in series, the total capacitance has _____ _____ (increased, decreased) to _____ _____ µF.

Experiment 39

INDUCTORS AND RL TIME CONSTANTS

SCHEMATIC:

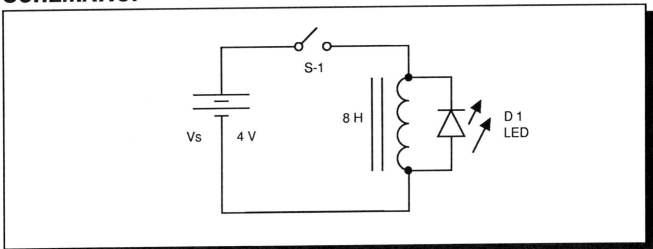

DISCUSSION

An inductor, which is just a coil of wire wound around some type of core, such as iron, can be used in the same way that a capacitor is used, for making a timing circuit. The RL timing circuit is not used as much as the RC circuit, however, because of the problems associated with the use of coils. First of all, the coil radiates a magnetic field, which can be a source of problems for circuitry that may be close by. Secondly, the RL circuit tends to cause an arc when you attempt to disconnect the circuit. And third, the coil tends to be heavier than a capacitor. In spite of all these problems, however, the coil is still used in many applications, and its operation must be understood by the technician.

TEXT CORRELATION

Before beginning this experiment, review section 15–4 on RL time constants in **Electricity.**

OBJECTIVES

1. To connect a resistor and inductor in a DC series circuit.

2. To analyze the operation of the inductive resistive DC circuit.

EQUIPMENT

- ☐ Standard tool box
- ☐ Standard parts kit
- ☐ Digital multimeter, with 10 megohm input resistance on voltage scales AND
- ☐ Volt-ohm-milliammeter, 20 kilohm/volt DC
- ☐ Adjustable power supply, capable of 0 to 40 VDC
- ☐ One coil, approximately 8 H in size, with no more than 250 ohms resistance
- ☐ One light emitting diode (LED), any visible color, 30 mA rating or higher

153

154 ■ Inductors and RL Time Constants

Experiment 39

PROCEDURES	FINDINGS	CONCLUSIONS
1. Using the multimeter set up as an ohmmeter, measure the DC resistance of the 8 H coil. Record the results in the blank indicated.	1. Record the reading obtained in the blank below: _____ohms	1. What causes the resistance of the coil to be so high? _____ _____ _____ _____.
2. Calculate the value of one time constant, using the formula $1\tau = L/R$ (use the resistance of the coil for the value of R in the formula). Record the calculation.	2. Record the calculated value in the blank below: _____seconds	—
3. Connect the circuit shown in Schematic 1. Be careful to connect the LED with the cathode end toward the + side of the power supply and the anode end toward the - side of the power supply. A simple way to test this is to energize the power supply and close the S_1 switch. If the LED comes on and stays on, IMMEDIATELY deenergize the power supply and reverse the leads of the LED. Otherwise, the LED may be damaged.	—	—
4. Energize the power supply. The LED should not be lit.	4. Observe the condition of the LED. It is _____ (lit/not lit).	4. Caution: If the LED comes on and stays on, deenergize the circuit immediately, and reverse the leads of the LED.
5. Open switch S_1, while observing the LED. Record the results.	5. Observe the condition of the LED. What did it do? _____ _____ _____ _____.	5. Why did the LED flash on, then go back off? _____ _____ _____ _____.
6. Increase the power supply voltage to 10 volts and observe and record the results, while performing steps 4 and 5.	6. Observe the condition of the LED. What did it do? _____ _____ _____ _____.	6. What caused the LED to flash brighter when the power supply voltage was increased? _____ _____ _____ _____.

Experiment 40

OPERATION OF DC RELAYS

SCHEMATIC:

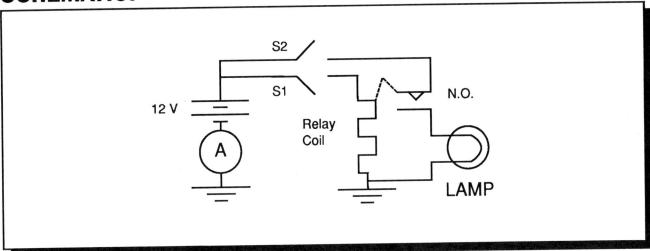

DISCUSSION

Relays are sometimes used to control circuitry from a remote location, or to control high voltage or high current circuits using low voltages for the control circuits. A relay is an electromechanical device, that, when current flows through it (through the relay coil), will produce a magnetic field strong enough to pull in the contacts of a switch located in the magnetic field of the relay coil. Some relays have only one set of contacts, which may be either normally open or normally closed, while other relays may have many sets of contacts to control several circuits at once.

TEXT CORRELATION

Before beginning this experiment, review section 8–9 on relays in *Electricity*.

OBJECTIVES

1. To test a DC relay for proper operation.

2. To measure the relay to determine resistance of the coil, resistance of the contacts, and current necessary to energize the relay and hold it energized.

EQUIPMENT

- ☐ Standard tool box
- ☐ Standard parts kit
- ☐ Digital multimeter, with 10 megohm input resistance on voltage scales OR
- ☐ Volt-ohm-milliammeter, 20 kilohm/volt DC
- ☐ Adjustable power supply, capable of 0 to 40 VDC output
- ☐ 1 DC relay, archer #275-206 or equivalent, rated at 12 VDC and 75 mA
- ☐ 1 #1004, #1157, or equivalent 12 volt automotive light bulb

155

PROCEDURES	FINDINGS	CONCLUSIONS
1. Locate the ON/OFF switch on the front panel of the power supply. Place it in the OFF position.	—	—
2. Locate the voltage adjust control on the front panel. Rotate the control fully CCW (counterclockwise).	2. This should adjust the power supply for an output of 0 volts.	
3. Connect the circuit shown in Schematic 1. Both S_1 and S_2 should be in the OFF position.	—	—
4. Close S_1. Energize the power supply. Slowly adjust the output of the power supply, while observing the operation of the relay and the reading on the ammeter. At the point where the relay energizes, evidenced by a "click" sound from the relay (if the case of the relay is a clear plastic, you may be able to see the movement of the relay contactor), stop adjusting the power supply. Record the current reading on the ammeter.	4. Record the pull in current of the relay. _____	4. Why does the relay require so much more current to pull in the relay than it does to keep the relay energized? _____ _____ _____ _____
5. Slowly adjust the power supply voltage in a downward direction, while observing the operation of the relay, and the ammeter reading. At the point where the relay deenergizes, evidenced again by the 'click' sound from the relay, stop adjusting the power supply.	5. Record the drop out current of the relay. _____	—
6. Remove the meter from the circuit. Readjust the meter to measure resistance, and place it on the lowest resistance scale.	—	—

Experiment 40 Operation of DC Relays ■ 157

PROCEDURES	**FINDINGS**	**CONCLUSIONS**
7. Readjust the power supply to read 12 volts (or whatever the relay is rated at). The relay should reenergize. Measure the resistance across the "normally open" contacts of the relay.	7. Record the resistance of the "normally open" contacts, with the relay energized. _____	7. Keep in mind that the test leads of the meter have a small amount of resistance. If the resistance of the test leads is taken into account, how much resistance the normally open contacts have, when the relay is energized? _____ _____ _____ _____.
8. Open switch S_1, deenergizing the relay. Observe the reading on the ohmmeter.	8. Record the resistance of the "normally open" contacts, with the relay deenergized.	
9. With S_1 still in the OFF position, close S_2. Observe the condition of the lamp.	9. Record whether the lamp is ON or OFF. _____	
10. Close S_1. Observe the condition of the lamp.	10. Record whether the lamp is ON or OFF. _____	10. Did the light illuminate when the relay closed? _____ Why would we use a relay to light the lamp? _____ _____ _____ _____.

Experiment 41
A.C. VOLTAGE

SCHEMATIC:

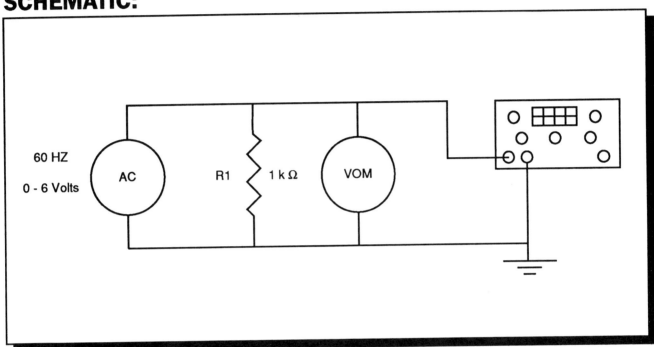

DISCUSSION

In addition to DC there exists a form of current known as alternating current (AC). AC derives its name from the fact that it periodically reverses (alternates) direction within the circuit. This results from the polarity of the source voltage constantly changing.

TEXT CORRELATION

Before beginning this experiment, review section 16–3 on magnitude of AC in *Electricity*.

OBJECTIVES

1. To use a volt-ohm-milliammeter (VOM) to measure V_{eff} and V_p.
2. To compare measured and calculated V_{eff} and V_p values.
3. To use the oscilloscope to measure the period and frequency of an AC sine wave.

EQUIPMENT

- ☐ Standard Tool Box
- ☐ Standard Parts Kit
- ☐ 0 to 6 volt isolated AC power source
- ☐ VOM
- ☐ Triggered sweep oscilloscope

PROCEDURES | FINDINGS | CONCLUSIONS

1. Connect the circuit shown in the schematic.   —   —

2. Set the VOM to the 10 volts AC range.   —   —

3. Check the calibration of the oscilloscope.   —   —

4. Adjust the vertical input control of the scope to 2 volts per division, the time per division control to 2 ms per division, and the AC coupled and internally triggered.   —   —

5. Starting at 0 volts, increase the voltage to the circuit in 1 volt increments.   —   —

6. Record in Table 1 the VOM readings and the number of vertical divisions from top to bottom (peak to peak) the waveform covers on the oscilloscope CRT as measured on R_1.

6. Table 1

VOM Volts	No. of Vert. Div. on CRT	Calculated Voltage
1		
2		
3		
4		
5		
6		

6. AC voltages can be measured by _____ and _____.

The oscilloscope measures AC voltages in terms of their _____ or _____ values.

How do the measurements taken from the scope and VOM compare? _____ _____ _____.

7. Sketch the image on the screen of the scope on Graph 1.

7. Graph 1

[Voltage vs. Time grid]

7. This waveform is known as a _____.

8. Using the conversion formula $V_{eff} = .3535 \times V_{p-p}$, calculate and record the voltage as taken from the scope.

8. V_{eff} (measured) _____
   V_{eff} (calculated) _____

8. Voltages measured as peak values can by converted to effective values by the formula _____.

Experiment 41 AC Voltage ■ 161

PROCEDURES	FINDINGS	CONCLUSIONS
		Do these calculated and measured voltages agree? _____.
9. From the CRT, count and record the number of divisions 1 cycle covers.	9. Number of horizontal divisions = _____ Horizontal sweep setting = _____ time/division.	9. This measurement is referred to as the _____.
10. Using the formula Frequency (f) = 1/period, calculate and record the f of the source.	10. Frequency = _____.	10. The oscilloscope has an advantage over the VOM as an AC voltage measuring device because it can also be used to determine _____.
11. Turn off your equipment.	—	—

Experiment 42

THE OSCILLOSCOPE:
Familiarization and Setup

DRAW THE FACE OF YOUR OSCILLOSCOPE HERE.

DISCUSSION

The oscilloscope (scope) is an electronic test instrument used to measure AC and DC voltages, frequencies, and phase angles. Unlike meters, it provides a television-like picture of electrical phenomena. These video images can provide detail of the waveshape that is otherwise unobtainable yet essential to a full understanding of many electrical circuits.

TEXT CORRELATION

Before beginning this experiment, review section 16–4 on AC time scale in *Electricity*.

OBJECTIVES

1. To become familiar with the oscilloscope controls.
2. To obtain a straight horizontal line on the vertical center of the oscilloscope screen (CRT).

EQUIPMENT

☐ Triggered sweep oscilloscope with test cables.

☐ Operator's manual for your oscilloscope

164 ■ The Oscilloscope   Experiment 42

SAFETY NOTES

  The CRT of the oscilloscope is constructed of a hollow GLASS vacuum tube. Be careful not to strike the face of this tube as an implosion causing serious injury may result.

PROCEDURES	**FINDINGS**	**CONCLUSIONS**

1. Examine the face of the scope, leads, and jacks. — —

2. Sketch a picture of the face of the scope in the space provided above. — —

3. Record the names of the various controls on the face of the scope.

3. Front Panel Control Names:
 A. _____
 B. _____
 C. _____
 D. _____
 E. _____
 F. _____
 G. _____
 H. _____
 I. _____
 J. _____
 K. _____
 L. _____
 M. _____
 N. _____
 O. _____
 P. _____
 Q. _____
 R. _____

3. Describe the function of each of these controls.
 A. _____
    _____
 B. _____
    _____
 C. _____
    _____
 D. _____
    _____
 E. _____
    _____
 F. _____
    _____
 G. _____
    _____
 H. _____
    _____
 I. _____
    _____

Experiment 42

The Oscilloscope ■ 165

PROCEDURES	FINDINGS	CONCLUSIONS
		J. _____
		K. _____
		L. _____
		M. _____
		N. _____
		O. _____
		P. _____
		Q. _____
		R. _____
4. Using the operator's manual, determine the function of each of the above controls.	—	—
5. Connect the leads to the scope.	—	—
6. Short the input leads (probes) together.	—	—
7. Set the trigger selector (source) switch to internal (INT) and the horizontal sweep (time per division) to external (EXT) or off.	—	—

PROCEDURES	FINDINGS	CONCLUSIONS
8. Turn on the scope and allow a one-minute warm-up time.	—	—
9. Adjust the brightness, focus, astigmatism (if present), and horizontal and vertical position controls to obtain a bright (but without halo) dot in the center of the screen.	—	—
10. Using the horizontal position control, move the dot to the first vertical line on the far left side of the screen.	—	—
11. Adjust the sweep selector to internal and slowly rotate the sweep speed (time per division) control to 1 ms.	11. What happened as the sweep speed was increased?	—
12. Alternately rotate the vertical position to the left (counter-clockwise) and right (clockwise). Do the same with the horizontal position control.	12. What happened as the vertical position control was rotated back and forth? What happened as the horizontal was rotated?	—
13. Turn off the scope.	—	—

Experiment 43

THE OSCILLOSCOPE:
DC Voltage Measurements

SCHEMATIC:

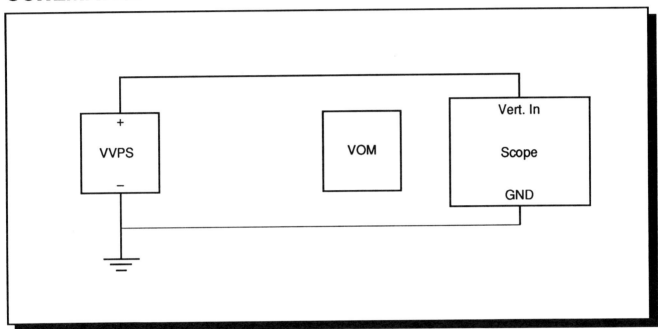

DISCUSSION

The scope is an instrument used to measure a variety of electrical phenomena. One of these is steady state (constant amplitude) DC voltage. In this experiment the scope will be used to measure this type of voltage.

TEXT CORRELATION

Before beginning this experiment, review section 16–5 on sine waves graphed vs. angle in degrees in *Electricity*.

OBJECTIVES

1. To use the scope to measure a variety of constant amplitude DC voltages.

2. To use the scope to measure reverse polarity DC voltages.

EQUIPMENT

- ☐ Standard Tool Kit
- ☐ Standard Parts Kit
- ☐ Triggered sweep oscilloscope with cables
- ☐ Variable voltage DC power supply (VVPS)
- ☐ VOM

SAFETY NOTES

 Remember! The CRT is a hollow GLASS vacuum tube. Be careful when handling.

PROCEDURES	FINDINGS	CONCLUSIONS
1. Wire the circuit shown in the schematic.	—	—
2. Obtain a sharp horizontal line on the vertical center of the CRT.	—	—
3. Be certain that the scope is in the vertical input (voltage) calibrated mode.	—	—
4. Adjust the scope for DC input coupling, its vertical input control (volts per division) to 2 volts per division, and its horizontal sweep (time per division) control to 1 millisecond per division.	—	—
5. Beginning at 0 volts, increase the voltage of the VVPS in 1 volt increments until 8 volts is reached as measured by the VOM.	—	—
6. Determine the voltage indicated by the scope by multiplying the number of divisions the line moved by the setting of the volts per division (vertical) control.	—	6. A scope can be used to measure a(n) _____ voltage. The amount of voltage can be determined by multiplying the setting of the _____ control by the number of _____ the horizontal line moved from its original zero volt position.

Experiment 43   The Oscilloscope ■ 169

PROCEDURES	FINDINGS	CONCLUSIONS

7. Record your findings in Table 1.

7. Table 1

Voltage from VOM	No. of Divisions	V from Scope
1		
2		
3		
4		
5		
6		
7		
8		

8. Return the VVPS voltage to 0.

9. Reverse the polarity of the VVPS connections to the circuit.

10. Reverse the connection of the VOM leads but leave the scope as originally wired.

10. The polarity of the voltage can be determined by observing the _____ in which the horizontal line moves.

11. Repeat step 6, recording the findings in Table 2.

11. Table 2

Voltage from VOM	No. of Divisions	V from Scope
1		
2		
3		
4		
5		
6		
7		
8		

12. Turn off the equipment.

Experiment 44

THE FUNCTION GENERATOR (SFG)

DRAW THE FACE OF YOUR FUNCTION GENERATOR HERE.

DISCUSSION

The function generator is an electronic test instrument used to produce AC voltages of various amplitudes, frequencies and wave-shapes. Typically sine, rectangular, and triangular sawtooth waveforms are available, with some generators producing ramp and pulse outputs as well. Some units are equipped with DC offset capabilities, which makes varying direct currents of the above wave shapes also available. In this experiment, we will attempt to become familiar with the front panel controls and hook-up of the basic function generator.

TEXT CORRELATION

Before beginning this experiment, review section 16–6 on AC alternators in *Electricity*.

OBJECTIVES

1. To become familiar with the function generator controls.

2. To obtain a variety of output waveforms and voltages from the function generator.

EQUIPMENT

- ☐ Function generator
- ☐ Operator's manual for the function generator
- ☐ Triggered sweep oscilloscope with test cables

172 ■ The Function Generator (SFG)   Experiment 44

SAFETY NOTES

 The CRT of the oscilloscope is constructed of a hollow GLASS vacuum tube. Be careful not to strike the face of this tube, as an implosion causing serious injury may result.

PROCEDURES	**FINDINGS**	**CONCLUSIONS**
1. Examine the front panel of the function generator, leads, and jacks.	—	—
2. Sketch a picture of the face of the generator in the space provided above.	—	—
3. Record the names of the various controls.	3. Front Panel Control Names: A. _____ B. _____ C. _____ D. _____ E. _____ F. _____ G. _____ H. _____ I. _____ J. _____ K. _____ L. _____	3. Describe the function of each of these controls. A. _____ _____. B. _____ _____. C. _____ _____ _____. D. _____ _____ _____. E. _____ _____ _____. F. _____ _____. G. _____ _____. H. _____

Experiment 44

The Function Generator (SFG)

PROCEDURES	FINDINGS	CONCLUSIONS

I. _____

J. _____

K. _____

L. _____

4. Connect the leads of the "high" output of the generator to the oscilloscope. — —

5. Turn on the scope and allow a one-minute warm-up time. — —

6. Adjust the frequency controls to 1 kHz, the waveform selector to sine wave, and rotate the level (amplitude) control from zero to about one-third of maximum. — —

7. Record the voltage as measured by the oscilloscope.
   7. Peak to peak voltage = _____. —

8. Increase the level control setting to about two-third's of maximum and record the new reading.
   8. Peak to peak voltage = _____.
   8. As the level control was rotated in the clockwise direction, the voltage got _____.

9. Leaving the level control set at two-third's, rotate the frequency control between 1 kHz and 5 kHz while observing the screen of the scope.
   9. What happened to the scope display as you rotated the frequency control? _____.
   9. As the frequency was increased, the number of cycles being displayed on the scope screen _____.

PROCEDURES	FINDINGS	CONCLUSIONS
		If the frequency was continuously increased, the sweep speed of the scope might have to be _____.
10. Turn off the equipment.	—	—

Experiment 45

OSCILLOSCOPE WITH NON-SINE WAVE INPUTS

SCHEMATIC:

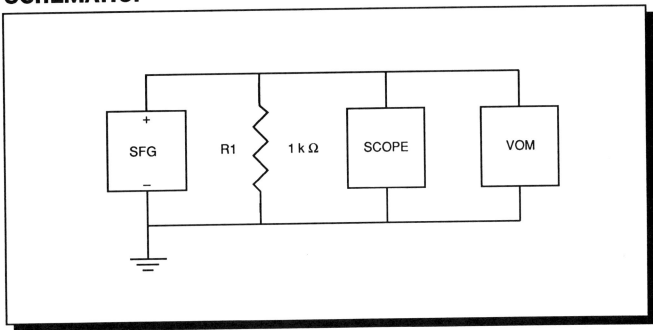

DISCUSSION

The scope is an instrument that is used to measure a variety of electrical phenomena. Among the most important of these is the shape of the waveforms themselves. In this experiment we will examine various waveforms produced by the function generator (SFG). In actual troubleshooting procedures, the circuit under test would itself be the source of the waveforms.

TEXT CORRELATION

Before beginning this experiment, review section 16–2 on Other AC wave forms in *Electricity*.

OBJECTIVES

1. To use the scope to display a variety of different wave shapes.

2. To use the scope to measure the frequency and amplitude of various wave shapes.

EQUIPMENT

- ☐ Standard tool kit
- ☐ Standard parts kit
- ☐ Triggered sweep oscilloscope with cables
- ☐ Function generator
- ☐ VOM

176 ■ The Oscilloscope with Non-Sine Wave Inputs

Experiment 45

SAFETY NOTES

 Remember! The screen of the scope is GLASS. Be careful not to strike it as an implosion may result.

PROCEDURES	**FINDINGS**	**CONCLUSIONS**

1. Wire the circuit shown in the schematic. — —

2. Obtain a sharp horizontal line on the vertical center of the CRT. — —

3. Check to be certain that the scope is in the vertical input (voltage) calibrated mode. — —

4. Adjust the scope for AC input coupling, the vertical input control (volts per division) for 2 volts per division and the horizontal sweep (time per division) control to 1 millisecond per division. — —

5. Turn on the function generator and set it to the rectangular wave (square wave) output waveform at a frequency of 100 Hz and a voltage of $4V_{p-p}$. — —

6. Draw the pattern appearing on the scope on Graph 1.

   6. Graph 1

   6. Does the waveform on the scope agree with the waveform setting indicated on the function generator?

   _____
   _____.

7. Increase the output frequency of the SFG in 100 Hz steps until 1 kHz is reached. —

   7. How do the VOM readings and SFG settings compare?

   _____
   _____
   _____
   _____.

Experiment 45

PROCEDURES	FINDINGS	CONCLUSIONS
8. Record your findings in Table 1.	8. Table 1 — **Rectangular Wave**	—

Freq. (Hz)	Number of Horizontal Divisions
100	
200	
300	
400	
500	
600	
700	
800	
900	
1000	

PROCEDURES	FINDINGS	CONCLUSIONS
9. Return the SFG to 0 volts.	—	—
10. Increase the frequency in 100 Hz steps and repeat steps 6 through 8.	—	10. The scope can be used to measure waveforms of various _____ _____.
11. Set the Function Generator to the triangular waveform function and while recording in Table 2, repeat steps 6 through 9. Draw the pattern appearing on the scope at 100 Hz on Graph 2.	11. Table 2 — **Triangular Wave**	—

Freq. (Hz)	Number of Horizontal Divisions
100	
200	
300	
400	
500	
600	
700	
800	
900	
1000	

PROCEDURES | FINDINGS | CONCLUSIONS

Graph 2

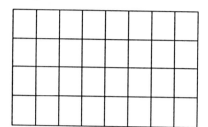

12. Turn off the equipment. | — | —

13. For further investigations: It is reletively easy to obtain a display on the scope at higher frequencies. It is often more difficult to obtain a stable display with slower signals. Set the signal generator on a frequency of 7 Hz. Try to obtain a stable display. It will probably be necessary to set the scope for normal triggering and to carefeully adjust the trigger LEVEL control. | — | —

Experiment 46

TRANSFORMERS

SCHEMATIC:

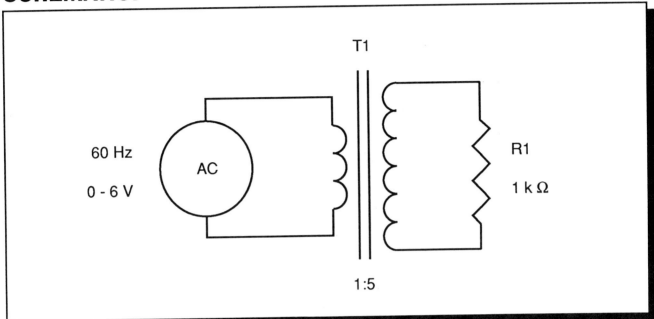

DISCUSSION

The transformer is a device constructed of coils of electrical conductive wire. It has the capability of increasing (stepping up) or decreasing (stepping down) AC voltages or currents, as well as being used as an AC coupling and isolation device or an ohmic (impedance) matching component. In this experiment we will examine its input (primary) and output (secondary) voltage and current characteristics.

TEXT CORRELATION

Before beginning this experiment, review section 17–2 on transformer voltage in *Electricity.*

OBJECTIVE

1. To determine the relationship of a transformer's turns ratio to its primary and secondary voltages and currents.
2. To demonstate how voltages can be stepped up or down.

EQUIPMENT

- [ ] Standard tool kit
- [ ] Standard parts kit
- [ ] Variable voltage 60 Hz power supply
- [ ] VOM
- [ ] Transformer specification sheet

SAFETY NOTES

A transformer is an inductive device. As with all such devices, high transient voltages can be generated by the transformer when it is turned off. This is sometimes referred to as "kick-back." Meters should be set to high ranges or removed before powering down this circuit.

PROCEDURES	FINDINGS	CONCLUSIONS
1. Using the VOM, measure the resistance of the primary and secondary coils of the transformer.	1. Primary resistance = _____. Secondary resistance = _____.	1. Is their any significant difference between these measurements? _____ If so, why? _____ _____ _____ _____ Do you think these measurements can be useful? _____ How? _____ _____ _____
2. Using the transformer specification sheet, determine the number of turns (loops of wire) on both the primary and secondary coils. With this information calculate the turns ratio (N_p/N_s).	2. Primary turns (N_p) = _____. Secondary turns (N_s) = _____. Turns ratio = _____.	2. This transformer is a step _____ transformer.
3. Wire the circuit shown in the schematic.	—	—
4. Adjust the variable AC power supply to 5 volts.	—	—
5. Using the VOM, measure the output (secondary) voltage of the transformer.	5. Primary voltage = _____. Secondary voltage = _____.	—
6. With the measured values of primary and secondary voltage, calculate the voltage ratio (V_p/V_s).	6. Voltage ratio = _____.	6. Compare this ratio to the turns ratio determined in step 2. How do they compare? _____ _____ _____

Experiment 46

Transformers ■ 181

PROCEDURES	FINDINGS	CONCLUSIONS
		What are your conclusions about these ratios? _____ _____ _____ _____.
7. Reverse the transformer (use the primary as the secondary and the secondary as the primary).	7. Caution: As a general practice this is not recommended as damage to the transformer can result. Check with your instructor.	—
8. Recalculate the turns ratio with this reversal.	8. Turns ratio = _____.	8. The transformer is now being used as a voltage step-_____ device.
9. Repeat steps 4 through 7.	9. Primary voltage = _____. Secondary voltage = _____. Voltage ratio = _____.	9. How do the turns ratio and voltage ratios now compare? _____ _____ _____.
10. Turn off the equipment.	—	10. What would happen to the output of the transformer if the input were a DC voltage? _____. Why? _____ _____ _____.

Experiment 47

TRANSFORMER – AC COUPLING

SCHEMATIC:

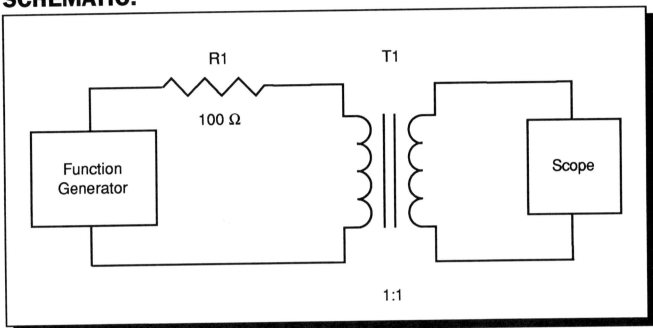

DISCUSSION

The transformer is a device constructed of coils of electrically conductive wire. It has the capability of increasing (stepping up) or decreasing (stepping down) AC voltages or currents. It may also be used as an AC coupling and isolation device or an ohmic (impedance) matching component. In this experiment, we will examine its capability to block DC voltages while passing (coupling) AC voltages.

TEXT CORRELATION

Before beginning this experiment, review section 17–8 on non-ideal transformers in *Electricity*.

OBJECTIVE

1. To use the transformer as an AC coupling device.

EQUIPMENT

- [ ] Standard tool kit
- [ ] Standard parts kit
- [ ] Function generator
- [ ] VOM
- [ ] Triggered sweep oscilloscope with cables

184 ■ The Transformer - AC Coupling
Experiment 47

SAFETY NOTES

 Be cautious of transient (kick-back) voltages generated by this circuit while turning off. Meters should be removed or set to high ranges while so doing.

PROCEDURES	FINDINGS	CONCLUSIONS

1. Wire the circuit shown in the schematic. — —

2. Adjust the function generator to 5 volts. — —

3. Turn on the DC offset of the generator and adjust it to 2 volts DC. — —

4. With the oscilloscope set for DC coupling, examine and compare the primary and secondary waveforms.

    4. Primary Waveform

    4. Describe the differences between the primary and secondary waveforms. _____
    _____
    _____
    _____.

    Secondary Waveform

5. Using the VOM, measure the output (secondary) voltage of the transformer.

    5. Primary voltage = _____.
    Secondary voltage = _____.

    5. How did the primary and secondary voltages compare?
    _____
    _____
    _____.

6. With the measured values of primary and secondary voltage, calculate the voltage ratio (V_p/V_s).

    6. Voltage ratio = _____.
    Turns ratio = _____.

    6. The voltage ratio of a transformer is the same as its _____.

    A transformer with this ratio is sometimes referred to as a(n) _____ transformer.

7. Turn off the equipment. — —

Experiment 48

INDUCTIVE REACTANCE

SCHEMATIC:

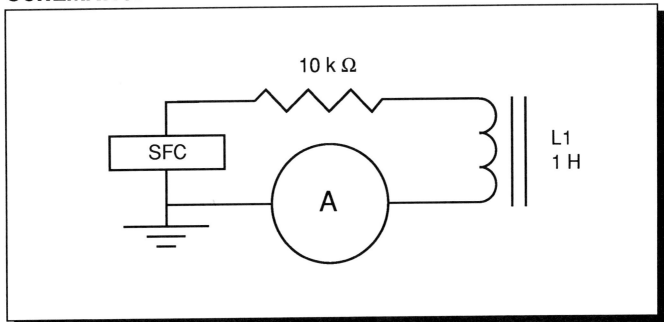

DISCUSSION

When a length of wire is formed into a coil (inductor) it takes on new and useful characteristics. One of the more notable of these is a form of opposition to current flow known as reactance (X). An investigation of this property will be undertaken in this laboratory procedure.

TEXT CORRELATION

Before beginning this experiment, review section 18–5 on inductive reactance in **Electricity**.

OBJECTIVES

1. To examine the relationship of inductive reactance (X_L) and inductance (L).

2. To determine the relationship between X_L and frequency (f).

3. To examine the relationship of X_L and voltage drops.

EQUIPMENT

- ☐ Standard tool box
- ☐ Standard parts kit
- ☐ Two VOM's
- ☐ Signal/function generator (SFG)

186 ■ Inductive Reactance

Experiment 48

SAFETY NOTES

 Be cautious of transient (kick-back) voltages that may be generated by the coil when powering down this circuit. Meters should be set to high ranges before turning off this circuit.

PROCEDURES	FINDINGS	CONCLUSIONS

1. Wire the circuit shown in the schematic.

   —   —

2. Adjust the output of SFG to a frequency of 100 Hz at an amplitude of 4 V_{eff}.

3. Record in Table 1 the voltages across R_1 and L_1 and the current through the circuit.

   3. Table 1

Freq. (Hz)	Volts R1	Volts L1	I meas.	I cal
100				
200				
300				
400				
500				
600				
700				
800				
900				
1000				

   —

4. Using the variation of Ohm's Law (I=V/X), calculate and record in Table 1 the total current.

   4. Table 2

Freq. (Hz)	Volts R1	Volts L1	I meas.	I cal
100				
200				
300				
400				
500				
600				
700				
800				
900				
1000				

   4. The _____ and _____ values are approximately the same. If they are not, try to explain the difference.

   In an AC circuit X_L acts similar to _____.

5. Vary the frequency of the SFG in 100 Hz steps up to 1000 Hz (1 kHz) while repeating procedures 3 and 4 for each of the settings.

   —

   5. As frequency increased X_L _____. This was demonstrated by the fact that the voltage _____.

PROCEDURES	FINDINGS	CONCLUSIONS
6. Return the output of the SFG to 0 volts.	—	—
7. Replace the 1.0 H inductor in the schematic with one of a 1.5 to 2.0 H value.	—	—
8. While recording the results in table 2, repeat procedures 2 through 6.	—	8. As inductance increases inductive reactance (X_L) _____. Check your graph before answering this question.
9. Turn off your circuit.	—	—
10. Plot the frequency and X_L data of Tables 1 and 2 on Graph 1.	10. Graph 1 X_L Frequency	—

Experiment 49

INDUCTIVE VOLTAGE DIVIDER

SCHEMATIC:

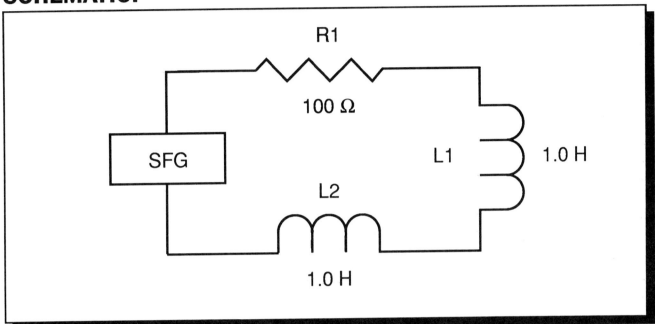

DISCUSSION

Inductors (coils) may be connected to form a basic series circuit. As with any series circuit, a voltage divider is created. The inductor is a reactive component, which means that it will exhibit an increased opposition to current when connected to an AC source. This makes inductors useful as AC voltage dividers. The ratio of voltage division is determined by the values (Henrys) of the inductors. Inductors typically have too few DC ohms, however, to be useful as DC voltage dividers. In this procedure, the use of inductors as AC voltage dividers will be examined.

TEXT CORRELATION

Before beginning this experiment, review section 18–6 on Ohm's Law for inductors in *Electricity*.

OBJECTIVES

1. To investigate the use of inductors as AC voltage dividers.
2. To determine the relationship between inductor values and voltage division.

EQUIPMENT

- ☐ Standard tool kit
- ☐ Standard parts kit
- ☐ VOM
- ☐ Signal or function generator (SFG)

SAFETY NOTES

 Remember! Inductors can create hazardous transient voltages when power to them is removed.

PROCEDURES	**FINDINGS**	**CONCLUSIONS**
1. Wire the circuit shown in the schematic.	—	—
2. Adjust the SFG to 5 volts effective at a frequency of 100 Hz.	—	—
3. Using the VOM, measure and record the voltages across each of the inductors.	3. Inductor 1 voltage = _____. Inductor 2 voltage = _____.	3. Does the voltage divide between the two components? _____. In an inductive AC voltage divider the larger inductor will have the _____ voltage. The property of inductors that allows them to be used as voltage dividers is called _____. The larger inductor has the _____ reactance and therefore has the _____ voltage.
4. Return the SFG to 0 volts and replace L_1 with a 1.5 H to 2.0 H inductor.	—	—
5. Repeat steps 2 and 3 with the new value of inductance.	5. Inductor 1 voltage = _____. Inductor 2 voltage = _____.	5. Increasing the value of L_1 caused its voltage to _____ and the voltage on L_2 to _____. Is this what you expected? _____ Explain. _____
6. Turn off the equipment.	—	—

Experiment 50

SERIES RL CIRCUITS WITH CHANGES IN INDUCTANCE

SCHEMATIC:

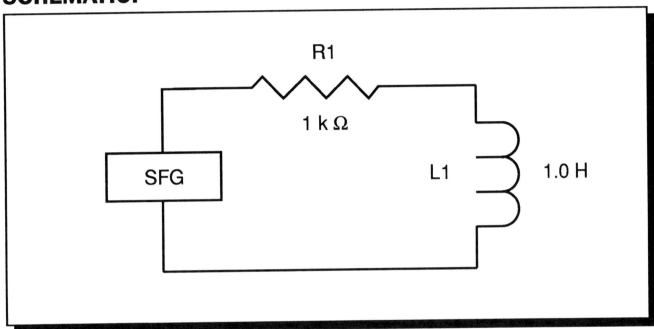

DISCUSSION

Resistive and reactive components are often combined in electrical circuits. When a coil (inductor) is connected in series with a resistor, a new form of opposition to current flow known as impedence (Z) is created. Unlike purely resistive ohms, these are affected by the inductance (L) of the coil and can cause the currents and voltages within the circuit to become "out of phase". This condition is indicated by a measurement known as phase angle. The current and voltage are "in phase" in the resistor, but not the inductor. The voltage leads the current by 90 electrical degrees.

TEXT CORRELATION

Before beginning this experiment, review section 19–6 on solving series AC circuits in *Electricity*.

OBJECTIVE

1. To examine the relationship of impedance (Z), inductive reactance (X_L), and phase angle to inductance (L) in a series RL circuit.

EQUIPMENT

☐ Standard tool box

☐ Standard parts kit

☐ Signal/function generator (SFG)

☐ VOM

192 ■ Series RL Circuits with Changes in Inductance Schematic

Experiment 50

SAFETY NOTES

 Remember! The inductor can produce destructive transient voltages when the circuit is being powered down.

PROCEDURES	**FINDINGS**	**CONCLUSIONS**

1. Wire the circuit shown in the schematic.

   —

   —

2. Adjust the output of the SFG to a frequency of 100 Hz and an amplitude of 4 V_{eff}.

   —

   —

3. Measure and record in Table 1 the voltages across R_1 and L_1.

   3. Table 1

L	VR1	VL	I	XL	Z	Angle
1.0 H						
2.0 H						
.5 H						

   —

4. Calculate and record the current, inductive reactance (X_L), impedance (Z), and phase angle.

   —

   4. Are the ratios of X_L to Z and V_L to V source approximately the same? _____.

   Should they be? _____.

   Explain why. _____

   _____

   _____

   _____

   _____.

   What formulas did you use to calculate:

   Inductive reactance? _____

   _____.

   Impedance? _____

   _____.

   Phase angle? _____

   _____.

   If the resistance of the circuit increased, the X_L would _____, the Z would _____, and the phase angle would

   _____.

Experiment 50   Series RL Circuits with Changes in Inductance Schematic ■ 193

PROCEDURES	FINDINGS	CONCLUSIONS
		If the resistance of the circuit decreased, the X_L would _____ _____, the Z would _____, and the phase angle would _____.
5. Replace L_1 with a 2.0 H coil and repeat steps 3 and 4.	—	5. As inductance increased X_L _____. This was demonstrated by the fact the voltage of the inductor _____ _____.
6. Replace L_1 with a .5 H coil and again repeat steps 3 and 4.	—	—
7. Turn off the circuit.	—	—
8. Calibrate Graph 1 and then plot on it the impedence (Z) and inductance data of Table 1.	8. Graph 1 Z 0.5 H 1.0 H 2.0 H Inductance	8. Check your graph while answering these questions. As inductance increased, the impedance _____. and the phase angle. Why? _____ _____ _____ _____. It can be concluded that the _____ _____, _____ and _____ of a series R_L circuit are directly related to the inductance of the coil.

Experiment 51

SERIES RL CIRCUITS WITH CHANGES IN FREQUENCY

SCHEMATIC:

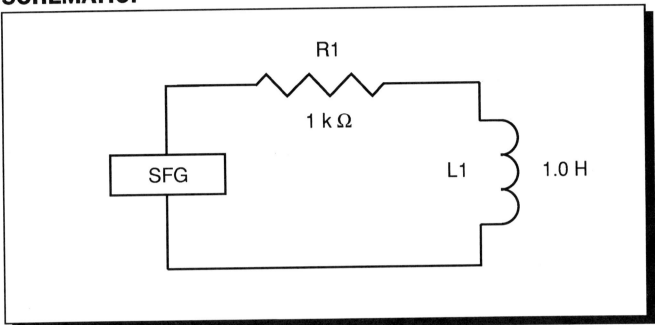

DISCUSSION

Resistive and reactive components are often combined in the same electrical circuit. When a coil (inductor) is connected in series with a resistor, a new form of opposition to current flow known as impedence (Z) is created. Unlike purely resistive ohms, these ohms are sensitive to the frequency of the source. These ohms can cause the currents and voltages in the circuit to become "out of phase". This condition is indicated by a measurement known as phase angle. In this procedure we shall examine how frequency is related to reactance, impedance, and phase angle in a series RL circuit.

TEXT CORRELATION

Before beginning this experiment, review section 19–6 on solving series AC circuits in *Electricity*.

OBJECTIVE

1. To determine the relationship of impedance (Z), inductive reactance (X_L) and phase angle to frequency (f) in a series RL circuit.

EQUIPMENT

- ☐ Standard tool box
- ☐ Standard parts kit
- ☐ Signal/function generator (SFG)
- ☐ VOM

196 ■ Series RL Circuits with Changes in Frequency   Experiment 51

SAFETY NOTES

 Remember! An inductor can generate destructive transient voltage when it is turned off. Be cautious when powering down this circuit.

PROCEDURES	**FINDINGS**	**CONCLUSIONS**

1. Wire the circuit shown in the schematic.   —   —

2. Adjust the output of the SFG to a frequency of 100 Hz and an amplitude of 4 V_{eff}.   —   —

3. Measure and record in Table 1 the voltages across R_1 and L_1.

3. Table 1

Freq.	VR1	VL	XL	Z	Angle
100					
200					
300					
400					
500					
600					
700					
800					
900					
1000					

—

4. Calculate and record the current, inductive ieactance (X_L), impedance (Z), and phase angle.

—

4. Are the ratios of X_L to Z and V_L to V source approximately the same?_____. Should they be?_____. Explain why._____
_____
_____
_____.

5. While repeating steps 3 and 4, increase the frequency of the SFG in 100 Hz increments until 1000 Hz (1 kHz) is reached.

—

5. As frequency increased X_L _____.

This was evidenced by the fact that the voltage of the inductor _____.

6. Turn off the circuit.   —   —

PROCEDURES	FINDINGS	CONCLUSIONS
7. Calibrate Graph 1 and plot the impedence (Z) vs. frequency data of Table 1.	7. Graph 1 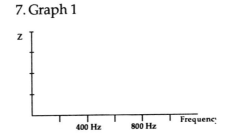	7. Check your graph while answering these questions. As frequency increased the impedence _____ and the phase. Explain why. _____ _____ _____. It can be concluded that the _____, _____ and _____ of a series RL circuit are directly related to the frequency of the source as all three _____ as the frequency increased.

Experiment 52

PARALLEL RL CIRCUITS WITH CHANGES IN INDUCTANCE

SCHEMATIC:

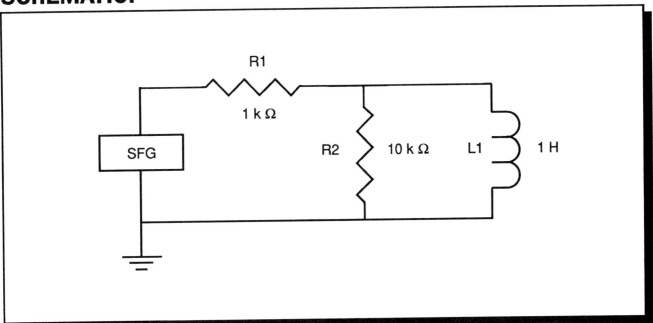

DISCUSSION

Reactive and resistive components may be connected in parallel as well as in series. In a parallel circuit, the current lags the voltage by 90°. As with any parallel circuit, a current divider is formed. In this procedure we will examine the current division, impedance, and phase angle of a parallel RL circuit and how they are affected by changes in the value of the inductor.

TEXT CORRELATION

Before beginning this experiment, review section 19-4 on solving parallel AC circuits in *Electricity*.

OBJECTIVES

1. To determine the impedance, phase angle, and currents in a parallel RL circuit.
2. To determine what effect changes in the value of L will have upon the parallel RL circuit.

EQUIPMENT

- ☐ Standard tool kit
- ☐ Standard parts kit
- ☐ Signal or function generator (SFG)
- ☐ VOM
- ☐ Triggered sweep oscilloscope

SAFETY NOTES

 While working with this circuit be aware that large transient voltages can result from inductive circuits when power is removed

PROCEDURES	FINDINGS	CONCLUSIONS
1. Wire the circuit shown in the schematic.	—	—
2. Adjust the SFG to 1 kHz at 5 volts.	—	—
3. Measure the source, R_1, R_2, and inductor voltages.	3. V source = _____. V resistor 1 = _____. V resistor 2 = _____. V inductor = _____.	—
4. Calculate the total, R_2, and inductor currents (I) and the inductive reactance (X_L).	4. I total = _____. I resistor 2 = _____. X_L = _____. I inductor = _____.	4. Does $I = I_L + I_{R2}$? _____ Do you think it should? _____ Why? _____ _____ _____ _____ _____
5. Let Z' (impedence of R_2/L_1) = V_{R2}/I total. Calculate Z'.	5. Z' = _____.	—
6. Using the formula I_L/I_{R2} calculate the phase angle of Z'.	6. Phase Angle = _____.	6. Is the sign of the angle negative or positive? _____ Is the current in the inductor leading or lagging the current in the resistor 2? _____ _____ _____
7. Replace the 1 H inductor with 1.5 H value and repeat steps 3 through 7.	7. V source = _____. V resistor 1 = _____. V resistor 2 = _____. V inductor = _____. I total = _____. I resistor 2 = _____. X_L = _____.	7. When the inductor was replaced with a different value did the following increase, decrease, or stay the same and explain why. _____ V resistor 1. _____ _____

Experiment 52   Parallel RL Circuits with Changes in Inductance ■ 201

PROCEDURES | FINDINGS | CONCLUSIONS

FINDINGS:
I inductor = _____.
Z' = _____.
Phase angle = _____.

CONCLUSIONS:
V resistor 2. _____
_____.
V inductor. _____
_____.
I total. _____
_____.
I resistor 2. _____
_____.
X_L. _____
_____.
I inductor. _____
Z'. _____
_____
_____.
Phase Angle. _____
_____
_____.

8. Turn off the equipment. —

8. If the inductor were open, what would happen to each?

a.) the total current in the circuit?

b.) the phase angle between the generator voltage and current?

c.) the generator voltage?

Experiment 53

PARALLEL RL CIRCUITS WITH CHANGES IN FREQUENCY

SCHEMATIC:

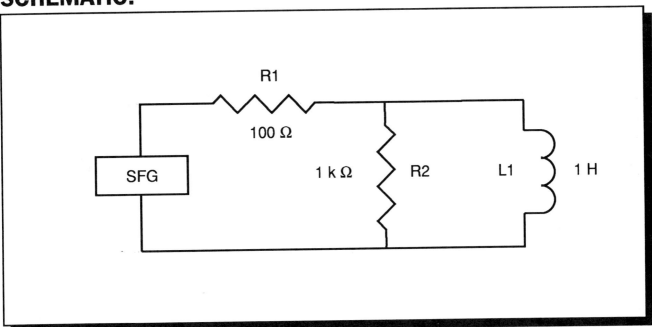

DISCUSSION

Reactive components such as inductors can be connected in parallel with resistive devices. In a parallel circuit, the current lags the voltage by 90°. As with any parallel circuit, a current divider is formed. In this experiment we will examine how current division, impedance, and phase angle in parallel RL circuits are affected by changes in the frequency of the source.

TEXT CORRELATION

Before beginning this experiment, review section 19–4 on solving parallel AC circuits in *Electricity*.

OBJECTIVES

1. To determine what effect changes in the value of source frequency will have on the impedance, phase angle, and currents in a parallel RL circuit.

EQUIPMENT

- ☐ Standard tool kit
- ☐ Standard parts kit
- ☐ VOM
- ☐ Signal or function generator (SFG)

204 ■ Parallel RL Circuits with Changes in Frequency

Experiment 53

SAFETY NOTES

 Remember! The inductor can produce potentially harmful transient voltages when the supply voltage is removed.

PROCEDURES	FINDINGS	CONCLUSIONS

1. Wire the circuit shown in the schematic. — —

2. Adjust the SFG to 100 Hz at 5 volts. — —

3. Measure the R_1, R_2, and inductor voltages.

3. Table 1

Freq. (Hz.)	VR1	VR2	VL	IR1	IR2	IL	Z'	Angle
100								
200								
300								
400								
500								
600								
700								
800								
900								
1000								
1100								
1200								
1300								
1400								
1500								

—

4. Calculate the total current (V_{R1}/R_1), the current through R_2 (V_{R2}/R_2), and I_L (V_L/X_L).

—

4. Does I total = $I_L + I_{R2}$? _____

How must the currents be added? _____

Why? _____
_____
_____
_____

Experiment 53 Parallel RL Circuits with Changes in Frequency ■ 205

PROCEDURES	FINDINGS	CONCLUSIONS
5. Let $Z' = V_{R2}/I$ total. Calculate Z'.	—	—
6. Using the expression tangent of the phase angle = "I_L/IR_2," calculate the phase angle.	—	6. Is the phase angle negative or positive? _____. Is the current in the inductor leading or lagging the current in R_2? _____.
7. Increase the frequency of the SFG in 100 Hz steps until 1.5 kHz is reached. Repeat steps 3 through 6 while recording the results in Table 1.	—	7. As the frequency was increased did the following increase, decrease, or stay the same? Explain why. _____ _____ _____ _____ _____. V resistor 1. _____ _____. V resistor 2. _____ _____. V inductor. _____ _____. I total. _____ _____. I resistor 2. _____ _____. I inductor. _____ _____. Z'. _____ _____. Phase angle. _____ _____.
8. Turn off the equipment.	—	—

Experiment 54

CAPACITIVE REACTANCE

SCHEMATIC:

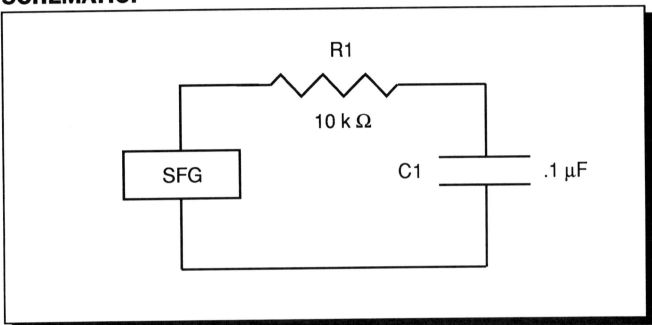

DISCUSSION

A capacitor is known as a reactive component. This means that capacitors have AC as well as DC ohmic properties. Unlike resistance, reactive ohms oppose only alternating current flow. These reactive ohms are affected by changes in capacitance as well as the frequency of the AC source. It should be noted that although the capacitor opposes AC, it does not pass steady-state DC at all. In this procedure we will examine capacitive reactance and its relationship to capacitance and the frequency of the source. The amount of current flow in the circuit is directly proportional to the capacitance.

TEXT CORRELATION

Before beginning this experiment, review section 18–2 on capacitive reactance in *Electricity*.

OBJECTIVES

1. To examine the relationship of capacitive reactance (X_C) to capacitance (C).
2. To examine the relationship of capacitive reactance (X_C) to frequency (f).

EQUIPMENT

- ☐ Standard tool box
- ☐ Standard parts kit
- ☐ Signal or function generator (SFG)
- ☐ VOM

208 ■ Capacitive Reactance Experiment 54

SAFETY NOTES

 Capacitors are capable of storing voltages for long periods of time after being removed from the source. Make certain capacitors are discharged before handling.

PROCEDURES	**FINDINGS**	**CONCLUSIONS**
1. Wire the circuit shown in the schematic.	—	—
2. Adjust the output of the SFG to a frequency of 100 Hz at an amplitude of 4 V_{eff}.	—	—
3. Record the voltages across R_1 and C_1.	3. Table 1	3. As frequency increased, X_c _____. The formula used to determine X_c is _____. Are your observations consistent with what the formula indicates should have happened? _____
4. Using the variation of Ohm's Law I=V/X, calculate and record the total circuit current.	—	—
5. Vary the frequency of the SFG in 100 Hz steps until 1000 Hz (1 kHz) is reached. Repeat procedures 3 and 4 for each of the frequency settings.	—	—
6. Return the output of the SFG to 0 volts.	—	—
7. Replace the .1 µF capacitor in the schematic with a .27 µF value.	—	—

Table 1

Freq. (Hz)	Voltage R1	Voltage C1	I Total
100			
200			
300			
400			
500			
600			
700			
800			
900			
1000			

Experiment 54 — Capacitive Reactance

PROCEDURES	FINDINGS	CONCLUSIONS
8. Recording in Table 2, repeat procedures 2 through 6.	8. Table 2	8. As capacitance increased, X_C _____. Is this consistent with your expectations? _____. Explain. _____.
9. Turn off your circuit.	—	—
10. Plot the frequency and X_C data of Tables 1 and 2 on Graph 1.	10. Graph 1 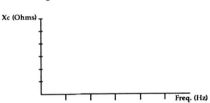	10. In an AC capacitive circuit, X_C acts similar to _____. Larger capacitors at any given frequency will produce a _____ value of X_C, while a given amount of capacitance will produce a _____ value of X_C as frequency is increased.

Table 2:

Freq. (Hz)	Voltage R1	Voltage C1	I Total
100			
200			
300			
400			
500			
600			
700			
800			
900			
1000			

Experiment 55

CAPACITIVE VOLTAGE DIVIDER

SCHEMATIC:

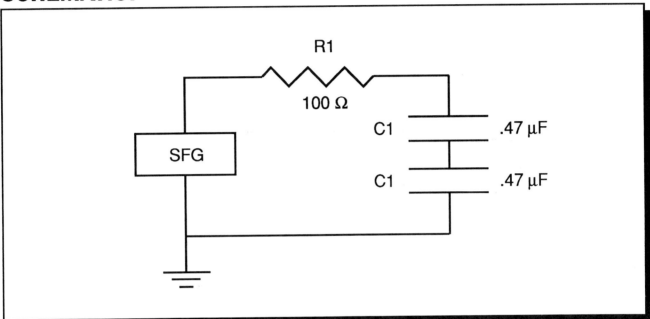

DISCUSSION

Like resistors, capacitors may be connected in series. Note that when this is done, the total capacitance actually decreases (capacitors in series add like resistors in parallel). As with any series circuit, a voltage divider is created. The capacitor is a reactive component, which means that, when connected to an AC source, it exhibits a change in opposition to current flow. Among its many applications is that of an AC voltage divider. The amount of voltage a capacitor will "drop" is related to its capacitance rating. In this procedure, the use of capacitors as AC voltage dividers will be examined.

TEXT CORRELATION

Before beginning this experiment, review section 18–3 on Ohm's Law for capacitors in *Electricity*.

OBJECTIVES

1. To examine the use of capacitors as AC voltage dividers.
2. To examine the relationship of capacitor ratings to voltage drops.

EQUIPMENT

- ☐ Standard tool kit
- ☐ Standard parts kit
- ☐ VOM
- ☐ Signal or function generator (SFG)

212 ■ Capacitive Voltage Divider   Experiment 55

PROCEDURES	**FINDINGS**	**CONCLUSIONS**
1. Wire the circuit shown in the schematic.	—	—
2. Calculate the total capacitance of the circuit.	2. When $C_1 = .47\ \mu F$, the total capacitance is _____. When $C_1 = .27\ \mu F$, the total capacitance is _____.	2. When capacitors are added in series the total capacitance _____.
3. Set the SFG to 100 Hz and 5 volts effective.	3. Capacitor 1 voltage = _____ Capacitor 2 voltage = _____	3. In a capacitive AC voltage divider the larger capacitor will have the _____ voltage drop. The property of capacitors that allow them to serve as voltage dividers is called _____. The larger capacitor has the _____ reactance and therefore has the _____ voltage.
4. Measure and record the voltage across each of the capacitors.	—	—
5. Return the voltage of SFG to 0 volts and replace C_1 with a $.27\ \mu F$ capacitor.	5. Capacitor 1 Voltage = _____ Capacitor 2 Voltage = _____	5. Increasing the value of C_1 caused the voltage on it to _____ and the voltage on C_2 to _____. Why? _____
6. Repeat procedures 2 and 3.	—	—
7. Turn off the circuit.	—	—

Experiment 56

SERIES RC CIRCUITS WITH CHANGES IN CAPACITANCE

SCHEMATIC:

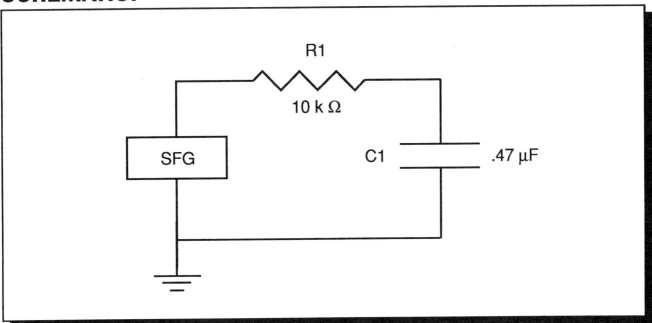

DISCUSSION

Resistive and reactive components are often combined in the same electrical circuit. When a capacitor is connected in series with a resistor, a new form of opposition to current flow known as impedance (Z) is created. Unlike purely resistive ohms, these are sensitive to the capacitance of the circuit and can cause the currents and voltages to be "out of phase". This condition is indicated by a measurement known as phase angle. In this procedure we will examine how capacitance affects the reactance, impedance and phase angle of a series RC circuit.

TEXT CORRELATION

Before beginning this experiment, review section 19–6 on solving series AC circuits in *Electricity*.

OBJECTIVES

1. To determine the relationship of capacitive reactance (X_c), impedance (Z) and phase angle to capacitance (C) in a series RC circuit.

EQUIPMENT

- ☐ Standard tool box
- ☐ Standard parts kit
- ☐ Signal or function generator (SFG)
- ☐ VOM

SAFETY NOTES

 Remember! A capacitor can store hazardous voltages. For this reason make certain they are discharged before handling.

PROCEDURES	**FINDINGS**	**CONCLUSIONS**

1. Wire the circuit shown in the schematic.   —   —

2. Adjust the output of SFG to a frequency of 100 Hz at an amplitude of 4 V_{eff}.   —   —

3. Measure and record in Table 1 the voltages across R_1 and C_1.

   3. Table 1

Freq.	VR1	VL	I	Xc	Z	Angle
.47 µF						
1.0 µF						
.27 µF						

   —

4. Calculate and record the current, capacitive reactance (X_c), impedance (Z), and phase angle.   —

   4. List the formulas used to calculate:

   Capacitive reactance = _____.

   Impedance = _____.

   Phase angle = _____.

   Calculate the V_c to V source and X_c to Z ratios. Voltage ratio = _____

   Ohms ratio = _____.
   Are these ratios approximately equal? _____. Should they be? _____.
   Explain. _____

Experiment 56   Series RC Circuits with Changes in Capacitance ■ 215

PROCEDURES	**FINDINGS**	**CONCLUSIONS**

5. Return the output of the SFG to 0 volts.

—

—

6. Replace the .47 µF capacitor in the schematic with a 1.0 µF value and repeat steps 2 through 4.

—

6. As capacitance increased, capacitive reactance _____ _____. This was evidenced by the fact that its voltage _____.

7. Replace the 1.0 µF capacitor with a .27 µF value and repeat steps 2 through 4.

—

—

8. Turn off your circuit.

—

—

9. Calibrate Graph 1, then plot the capacitance vs. impedance (Z) data of Table 1.

9. Graph 1

Z

.27 µF   .47 µF   1.0 µF
Capacitance

9. Graph 1 shows that as the capacitance decreased, the impedance of the circuit _____ _____ and as it increased, the Z _____.
Was this expected? _____

Why? _____

What effect do you think that increases in resistance will have on:

Capacitive reactance? _____

Impedence? _____

Phase angle? _____

What effect do you think that decreases in resistance will have on:

Capacitive reactance? _____

Impedence? _____

Phase angle? _____

Experiment 57

SERIES RC CIRCUITS WITH CHANGES IN FREQUENCY

SCHEMATIC:

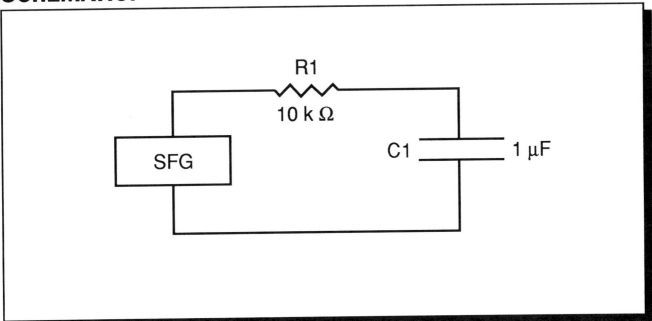

DISCUSSION

Resistive and reactive components are often combined in the same electrical circuit. When a capacitor is connected in series with a resistor, a new form of opposition to current flow known as impedance (Z) is created. Unlike purely resistive ohms these are sensitive to the frequency of the source and can cause the currents and voltages in the circuit to be "out of phase". This condition is indicated by a measurement known as phase angle. If the frequency is increased, more current will flow. This is why a capacitor is sometimes thought of as a high-frequency short.

TEXT CORRELATION

Before beginning this experiment, review section 19–6 on solving series AC circuits in *Electricity*.

OBJECTIVES

1. To determine the relationship of frequency (f) to the reactance (X_C), impedance (Z), and phase angle of a series RC circuit.

EQUIPMENT

- ☐ Standard tool box
- ☐ Standard parts kit
- ☐ Signal or function generator (SFG)
- ☐ VOM

SAFETY NOTES

 Remember! Capacitors can store hazardous voltages. Make certain they are discharged before handling.

PROCEDURES	FINDINGS	CONCLUSIONS

1. Wire the circuit shown in the schematic.

   —   —

2. Adjust the output of SFG to a frequency of 100 Hz at an amplitude of 4 V_{eff}.

   —   —

3. Measure and record in Table 1 the voltages across R_1 and C_1.

   3. Table 1

Freq.	VR1	Vc	I	Xc	Z	Angle
100						
200						
300						
400						
500						
600						
700						
800						
900						
1000						

   —

4. Calculate and record the current, capacitive reactance (X_c), impedance (Z), and phase angle.

   —

   4. List the formulas used to calculate:

   Capacitive reactance. _____

   _____.

   Impedance. _____

   _____.

   Phase angle. _____

   _____.

   Calculate the Vc to V source and Xc to Z ratios.

   Voltage ratio = _____.

   Ohms ratio = _____.

   Are these ratios approximately equal? _____.

   Should they be? _____.

   Explain. _____

   _____

   _____.

Experiment 57   Series RC Circuits with Changes in Frequency ■ 219

PROCEDURES	FINDINGS	CONCLUSIONS
5. While repeating steps 3 and 4, increase the frequency of the SFG in 100 Hz increments until 1000 Hz (1 kHz) is reached.	—	5. As frequency increased, capacitive reactance _____ _____. This was evidenced by the fact that its voltage _____.
6. Return the output of the SFG to 0 volts.	—	—
7. Turn off the circuit.	—	—

8. Calibrate then plot the frequency and Z data of Table 1 on Graph 1.	8. Graph 1
	Z (Ohms)
	                     Frequency (kHz)
	    1 kHz
		8. As the frequency increased, the impedence _____. Why? _____ _____ _____ _____.

Experiment 58

PARALLEL RC CIRCUITS WITH CHANGES IN CAPACITANCE

SCHEMATIC:

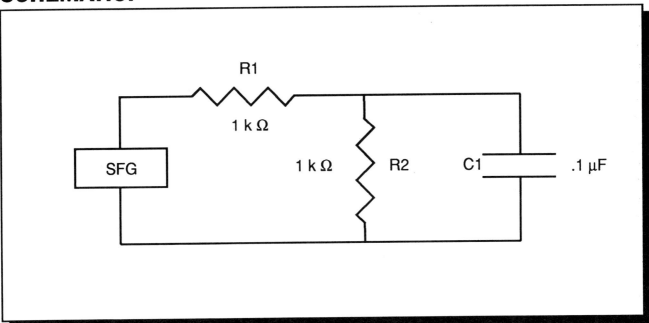

DISCUSSION

Reactive components such as capacitors can be connected in parallel with resistive devices. As with any parallel circuit, a current divider is formed. In this experiment we will examine the current division, impedance, and phase angle relationships found in parallel RC circuits. We will investigate how these relationships are affected by changes in the value of capacitance.

TEXT CORRELATION

Before beginning this experiment, review section 19-4 on solving parallel AC circuits in *Electricity*.

OBJECTIVE

1. To determine the impedance, phase angle, and currents in a parallel RC circuit and how they are affected by changes in capacitance.

EQUIPMENT

- ☐ Standard tool kit
- ☐ Standard parts kit
- ☐ VOM
- ☐ Signal or function generator (SFG)

SAFETY NOTES

Remember! The capacitor can store energy for a long period of time. Make certain they are discharged before handling. Remember! Some capacitors are polarized and cannot be used with AC sources as they can become hazardous. For this reason, only non-polarized capacitors may be used in this experiment.

PROCEDURES	FINDINGS	CONCLUSIONS
1. Wire the circuit shown in the schematic.	—	—
2. Adjust the SFG to 1 kHz at 5 volts.	—	—
3. Measure the source, R_1, R_2, and capacitor voltages.	3. V source = _____. V resistor 1 = _____. V resistor 2 = _____. V capacitor = _____.	—
4. Calculate the total current (V_{R1}/R_1), the current through R_2 (V_{R2}/R_2), X_c ($1/2(\pi)fC$), and I_c (Vc/Xc).	4. I total = _____. I resistor 2 = _____. X_c = _____. I capacitor = _____.	4. Does I total = $I_c + I_{R2}$? _____. Do you think it should? _____. Why? _____ _____ _____ _____
5. Let Z' = VR2/I total. Calculate Z'.	—	—
6. Using the expressions, "tangent of the phase angle = I_c/I_{R2}," calculate the phase angle.	6. Z' = _____.	—
7. Replace the .1 µF capacitor with a .27 µF value and repeat steps 3 through 6.	7. Phase Angle = _____.	—
8. Turn off the equipment.	8. V source = _____. V resistor 1 = _____. V resistor 2 = _____. V capacitor = _____. I total = _____. I resistor 2 = _____. X_c = _____.	8. When the capacitor was replaced did the following increase, decrease, or stay the same? Explain why. _____ V resistor 1. _____ V resistor 2. _____

Experiment 58
Parallel RC Circuits with Changes in Capacitance ■ 223

PROCEDURES	FINDINGS	CONCLUSIONS
	I capacitor = _____.	V capacitor. _____
	Z′ = _____.	_____
	Phase angle = _____.	I total. _____
		_____
		I resistor 2. _____
		_____
		X_C. _____
		_____
		I capacitor. _____
		_____
		Z′. _____
		_____
		Phase angle. _____
		_____

Experiment 59

PARALLEL RC CIRCUITS WITH CHANGES IN FREQUENCY

SCHEMATIC:

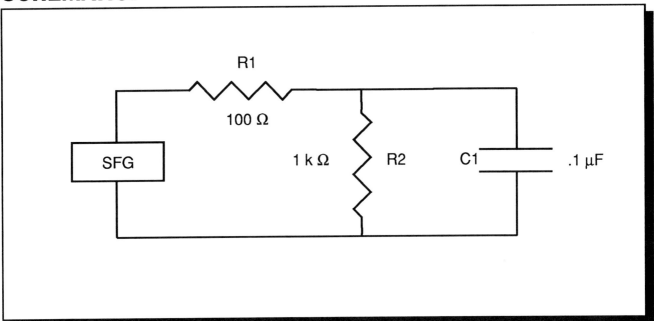

DISCUSSION

Reactive components such as capacitors can be connected in parallel with resistive devices. As with any parallel circuit, a current divider is formed. In this experiment we will examine how current division, impedance, and phase angle are affected by changes in the source frequency to parallel RC circuits.

TEXT CORRELATION

Before beginning this experiment, review section 19-4 on solving parallel AC circuits in *Electricity*.

OBJECTIVES

1. To determine what effect changes in the value of source frequency will have on the impedance, phase angle, and currents in a parallel RC circuit.

EQUIPMENT

- ☐ Standard tool kit
- ☐ Standard parts kit
- ☐ VOM
- ☐ Signal or function generator (SFG)

SAFETY NOTES

 Remember! Capacitors can store energy for long periods of time. Make certain they are discharged before handling. Remember! Some capacitors are polarized and cannot be used with AC sources as they can become hazardous. For this reason, only non-polarized capacitors may be used in this experiment.

PROCEDURES	FINDINGS	CONCLUSIONS
1. Wire the circuit shown in the schematic.	—	—
2. Adjust the SFG to 1 kHz at 5 volts.	—	—
3. Measure the R_1, R_2, and capacitor voltages.	3. See next page for Table 1.	—
4. Calculate the total current (V_{R1}/R_1), the current through $R_2 = (V_{R2}/R_2)$, and $I_c = (V_c/X_c)$.		4. Does I total = $I_c + I_{R2}$? _____ _____. How must the currents be added. _____ _____. Why? _____ _____ _____ _____.
5. Let $Z' = V_{R2}/I$ total. calculate Z'.	—	—
6. Using the expressions, "tangent of the phase angle = I_c/I_{R2}," calculate the phase angle.	—	—
7. Increase the frequency of the SFG in 100 Hz steps until 1.5 kHz is reached. Repeat steps 3 through 6 while recording the results in Table 1.		7. As the frequency was increased did the following increase, decrease, or stay the same? Explain why. _____ _____ _____. V resistor 1. _____ _____. V resistor 2. _____ _____. V capacitor. _____ _____.

Experiment 59 — Parallel RC Circuits with Changes in Frequency

PROCEDURES | FINDINGS | CONCLUSIONS

I total. _____

I resistor 2. _____

I capacitor. _____

Z'. _____

Phase angle. _____

8. Turn off the equipment.

Table 1

Freq.	VR1	VR2	Vc	IR1	IR2	Ic	Z	Angle
100								
200								
300								
400								
500								
600								
700								
800								
900								
1000								
1100								
1200								
1300								
1400								
1500								

Experiment 60

SERIES LRC CIRCUITS

SCHEMATIC:

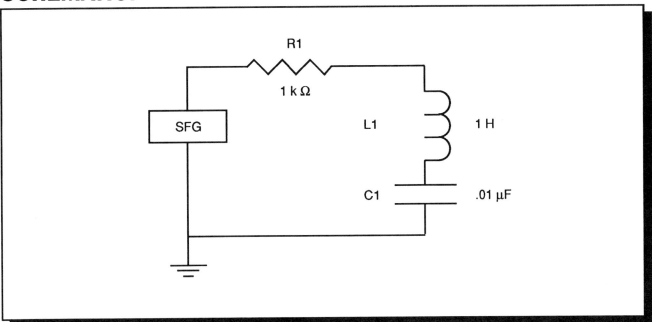

DISCUSSION

Capacitors and inductors possess a property known as reactance. As you know, this characteristic opposes AC current and is measured in units of ohms. Unlike resistive ohms, however, there exist ways in which these ohms can be reduced or, as we will discover in later experiments, eliminated completely. In this experiment we will take a look at the interaction of capacitive and inductive reactive ohms when connected in series.

TEXT CORRELATION

Before beginning this experiment, review section 20–1 on LRC series circuits in *Electricity*.

OBJECTIVES

1. To investigate the behavior of inductive and capacitive reactance when connected in a series circuit.

2. To determine how the phase angle of a circuit is affected when capacitors and inductors are connected in series.

EQUIPMENT

☐ Standard tool kit

☐ Standard parts kit

☐ VOM

☐ Signal or function generator (SFG)

230 ■ Series LRC Circuits   Experiment 60

SAFETY NOTES

Inductors and capacitors have energy capabilities. This energy can be a dangerously high amount. Also be alert to the high transient voltages that may be generated by inductors (coils) when power is removed.

PROCEDURES	FINDINGS	CONCLUSIONS
1. Wire the circuit shown in the schematic.	—	—
2. Adjust the SFG to 1 kHz at 5 volts.	—	—
3. Measure the source, R_1, inductor, and capacitor voltages.	3. V source = _____. V resistor 1 = _____. V inductor = _____. V capacitor = _____.	3. Does V total = $V_R + V_C + V_L$? Why not? _____ How must they be added? _____
4. Calculate the total current (V_{R1}/R_1), X_L, and X_C.	4. I total = _____. X_L = _____. X_C = _____.	—
5. Calculate the total impedance ($Z = V_t / I$ total) of the circuit.	5. Z total = _____.	—
6. Using the formula "tangent = $(X_L - X_C)/R$" calculate the phase angle of the circuit.	6. Phase angle = _____.	6. Is the sign of the angle negative or positive? _____. Is the voltage on the inductor leading or lagging the voltage on the resistor? _____. Is the capacitor voltage leading or lagging the circuit current? _____
7. Replace the 1 H inductor with 1.5 H value and repeat steps 3 through 7.	7. V source = _____. V resistor 1 = _____. V inductor = _____. V capacitor = _____. I total = _____. X_C = _____. X_L = _____. Z total = _____. Phase Angle = _____.	7. When the inductor was replaced with a different value did the following increase, decrease or stay the same? Explain why. _____ V resistor. _____ V inductor. _____

Experiment 60

PROCEDURES	FINDINGS	CONCLUSIONS

		V capacitor. _____
		I total. _____
		X_L. _____
		Z total. _____
		Phase angle. _____
8. Return the 1 H inductor to the circuit.	—	—
9. Replace the .01 µF with a .1 µF value and repeat steps 3 through 7.	9. V source = _____ V resistor 1 = _____ V inductor = _____ V capacitor = _____ I total = _____ X_C = _____ X_L = _____ Z total = _____ Phase angle = _____	9. When the capacitor was replaced with a different value did the following increase, decrease, or stay unchanged and explain why. _____
		V resistor. _____
		V inductor. _____
		V capacitor. _____
		I total. _____
		X_L. _____
		Z total. _____
		Phase angle. _____
10. Turn off the equipment.	—	—
11. Study the three sets of data.	—	11. Describe what happened and why to Z total and the phase angle when: The inductance was increased. _____

232 ■ Series LRC Circuits

Experiment 60

| **PROCEDURES** | **FINDINGS** | **CONCLUSIONS** |

_____
_____
_____

The capacitance was increased.

_____
_____
_____
_____
_____

Experiment 61

SERIES RESONANT CIRCUITS

SCHEMATIC:

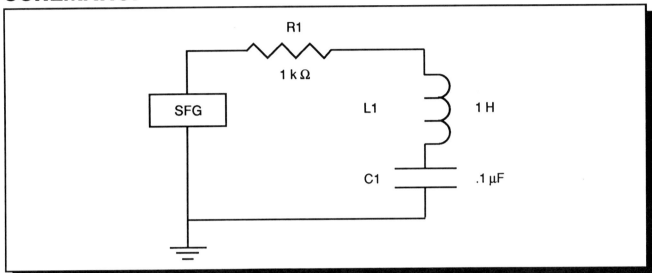

DISCUSSION

Capacitors and inductors possess a property known as reactance, which is measured in ohms. Unlike resistive ohms, however, these ohms are affected by the frequency of the applied AC voltage. The ohms of a capacitor and inductor operate in opposition to each other and can, at a specific frequency, become equal. In this experiment, we will take a look at the interaction of capacitive and inductive reactive ohms when connected in series and of equal value. This condition is known as resonance. At series resonance reactance phasors for inductors and capacitors cancel each other out.

TEXT CORRELATION

Before beginning this experiment, review section 20–1 on series resonant frequency in *Electricity*.

OBJECTIVES

1. To investigate the behavior of inductive and capacitive reactance when they are connected in series and equal value.

2. To determine how the phase angle of a circuit is affected when capacitors and inductors are connected in series and their reactances are equal in value.

3. To explain the factors affecting the selectivity of a series resonant circuit.

EQUIPMENT

- ☐ Standard tool kit
- ☐ Standard parts kit
- ☐ VOM
- ☐ Signal or function generator (SFG)

SAFETY NOTES

 Inductors and capacitors have energy storing capabilities. This energy is often of a high value and may be HAZARDOUS. Be cautious of the voltages stored by capacitors and the transient voltages created by inductors as the circuit is being powered down.

PROCEDURES	FINDINGS	CONCLUSIONS

1. Wire the circuit shown in the schematic. — —

2. Adjust the SFG to 100 Hz sine waves at 5 volts. — —

3. While measuring the voltage of R_1 (V_{R1}), increase in 100 Hz steps the frequency to the circuit until 1.5 kHz is reached.

   3. Table 1

   $C1 = .1 \, \mu F$

Freq. (Hz)	VR1	I tot.	VL1	Z	Angle
100					
200					
300					
400					
500					
600					
700					
800					
900					
1000					
1100					
1200					
1300					
1400					
1500					

4. Calculate the total current (V_{R1}/R_1) of the circuit at each of the frequencies. — —

5. Let Z = V total/I total. Calculate Z at each of the frequencies. — —

6. Using the formula "tangent of the phase angle = ($V_L - V_C)/V_{R2}$," calculate the phase angle of the circuit at each of the input frequencies. — —

7. Calculate the resonant Frequency (f_r) of the circuit.

   7. Resonant frequency = _____ Hz with $C1 = .1 \, \mu F$.

   Resonant frequency = _____ Hz with $C1 = .05 \, \mu F$.

   7. Did the resonant frequency change when the capacitor was replaced? _____. Did you think it would? ____

PROCEDURES	FINDINGS	CONCLUSIONS

Why?_____
_____
_____
_____
_____.

8. Replace the .1 µF capacitor with a .05 µF unit.

8. Table 2

C1 = .05 µF

Freq. (Hz)	VR1	I tot.	VL1	Z	Angle
100					
200					
300					
400					
500					
600					
700					
800					
900					
1000					
1100					
1200					
1300					
1400					
1500					

9. Repeat steps 3 through 7.

—

9. Was the total current, phase angle, and impedance greatest or least when the circuit was operated at (or near) the resonant frequency? _____.

Current: _____.
Why?_____
_____.

Phase angle: _____.
Why?_____
_____.

Z: _____.
Why?_____
_____.

Was the phase angle - or + when operating below f_r?
_____.

PROCEDURES	FINDINGS	CONCLUSIONS
		Was the phase angle − or + when operating above f_r? _____
10. Turn off the equipment.	—	—
11. Compare the two sets of data.	—	11. Did the current increase or decrease as you increased the SFG frequency to the circuit? _____
		Was it highest or lowest at resonance? _____
		Once the resonant frequency was exceeded, what happened to the current? _____

Experiment 62

PARALLEL LRC CIRCUITS

SCHEMATIC:

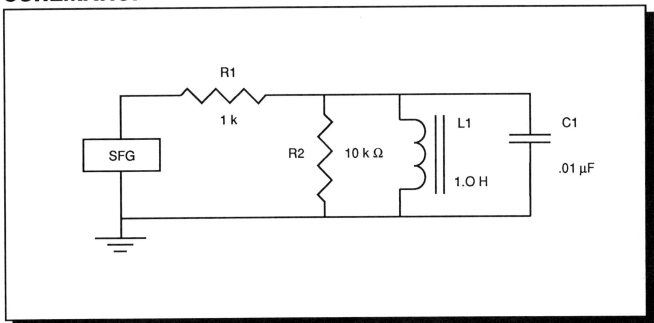

DISCUSSION

Capacitors and inductors possess a property known as reactance. As you know, this characteristic opposes AC current and is measured in units of ohms. Unlike resistive ohms, however, there exist ways in which these ohms can be reduced or, as we will discover later, eliminated completely. In this experiment we will take a look at the interaction of capacitive and inductive reactive ohms when connected in parallel.

TEXT CORRELATION

Before beginning this experiment, review section 20–7 on LRC parallel circuits in *Electricity*.

OBJECTIVES

1. To investigate the behavior of inductive and capacitive reactance when connected in a parallel circuit.

2. To determine how the phase angle of a circuit is affected when capacitors and inductors are connected in parallel.

EQUIPMENT

- ☐ Standard tool kit
- ☐ Standard parts kit
- ☐ Signal or function generator (SFG)
- ☐ Triggered sweep oscilloscope
- ☐ VOM

SAFETY NOTES

Inductors and capacitors can store energy and create transient voltages. Protect yourself and equipment against these during and after the circuit has been powered down.

PROCEDURES	FINDINGS	CONCLUSIONS
1. Wire the circuit shown in the schematic.	—	—
2. Adjust the SFG to 1 kHz at 5 volts.	—	—
3. Measure the source, R_1, R_2, inductor, and capacitor voltages.	3. V source = _____. V resistor 1 = _____. V resistor 2 = _____. V inductor = _____. V capacitor = _____.	
4. Calculate the total current [$I_t = V_{R1}/R_1$], the current through R_2 [$I_{R2} = V_{R2}/R_2$], inductive reactance [$X_L = 2\pi fL$], current of the inductor [$I_L = V_L/X_L$], capacitive reactance [$X_C = 1/2\pi fC$], and current through the capacitor [$I_C = V_C/X_C$].	4. I total = _____. I resistor 2 = _____. X_L = _____. X_C = _____. I capacitor = _____. I inductor = _____.	4. Does I total = $I_L + I_C + I_{R2}$? _____. Do you think it should? _____. Why? _____.
5. Calculate the total impedance ($Z = V_t/I$ total) of the circuit.	5. Z total = _____.	—
6. Let $Z' = V_{R2}/I$ total. Calculate Z'.	6. Z' = _____.	—
7. Using the formula "tangent of the phase angle = $I_C - I_L / I_{R2}$," calculate the phase angle of Z'.	7. Phase angle = _____.	7. Is the sign of the angle negative or positive? _____. Is the current in the inductor leading or lagging the current in the resistor 2? _____. Is the inductive current leading or lagging the current in resistor 2? _____.
8. Replace the 1 H inductor with 1.5 H value and repeat steps 3 through 7.	8. V source = _____. V resistor 1 = _____. V resistor 2 = _____.	8. When the inductor was replaced with a different value did the following increase, decrease, or stay the same?

Experiment 62 Parallel LRC Circuits ■ 239

PROCEDURES	FINDINGS	CONCLUSIONS
	V inductor = _____.	Explain why.
	V capacitor = _____.	V resistor 1. _____.
	I total = _____.	
	I resistor 2 = _____.	V resistor 2. _____.
	I capacitor = _____.	
	X_c = _____.	V inductor. _____.
	I inductor = _____.	
	X_L = _____.	V capacitor. _____.
	Z total = _____.	
	Z' = _____.	I total. _____.
	Phase angle = _____.	
		I resistor 2. _____.
		X_L. _____.
		I inductor. _____.
		I capacitor. _____.
		Z total. _____.
		Z'. _____.
		Phase angle. _____.
9. Return the 1 H inductor to the circuit.	—	—
10. Replace the .01 µF with a .1 µF value and repeat steps 3 through 7.	10. V source = _____.	10. When the capacitor was replaced with a different value did the following increase, decrease, or stay unchanged?
	V resistor 1 = _____	
	V resistor 2 = _____.	Explain why.
	V inductor = _____.	V resistor 1. _____.
	V capacitor = _____.	
	I total = _____.	V resistor 2 _____.
	I resistor 2 = _____.	

PROCEDURES | FINDINGS | CONCLUSIONS

I capacitor = _____.

X_C = _____.

I inductor = _____.

X_L = _____.

Z total = _____.

Z′ = _____.

Phase angle = _____.

V inductor. _____.

V capacitor. _____.

I total. _____.

I resistor 2. _____.

X_C _____.

I capacitor. _____.

Z total. _____.

Z′. _____.

Phase angle. _____.

11. Turn off the equipment.

12. Examine the three sets of data.

Experiment 63

PARALLEL RESONANT CIRCUITS

SCHEMATIC:

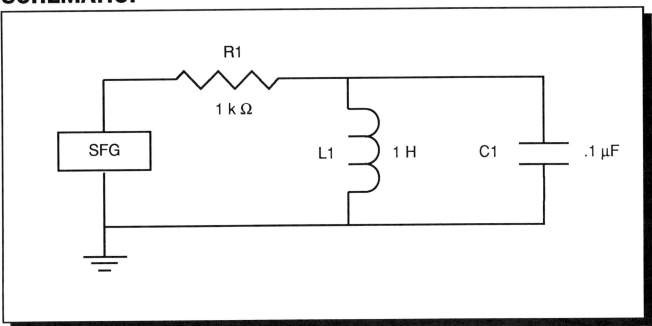

DISCUSSION

Capacitors and inductors possess a property known as reactance, which is measured in ohms. Unlike resistive ohms, however, these ohms are affected by the frequency of the applied AC voltage. The ohms of capacitors and inductors operate in opposition to each other and can, at a specific frequency, become equal. In this experiment, we will examine the interaction of reactive ohms when they are wired in parallel and equal in value, a condition known as parallel resonance. It should be found that the total current is at a minimum when the circuit is operated at the resonant frequency.

TEXT CORRELATION

Before beginning this experiment, review section 20–8 on parallel resonant frequency in *Electricity*.

OBJECTIVES

1. To investigate the behavior of inductive and capacitive reactance and phase angle when connected in parallel and of equal value.
2. To measure the frequency response of a parralel resonant circuit.

EQUIPMENT

- ☐ Standard tool kit
- ☐ Standard parts kit
- ☐ VOM
- ☐ Signal or function generator (SFG)

242 ■ Parallel Resonant Circuits | Experiment 63

SAFETY NOTES

 Inductors and capacitors have energy storing capabilities. This energy is often of a high value and may be HAZARDOUS. Be cautious of transient voltages created by inductors when power is removed and the voltages stored by capacitors.

PROCEDURES	FINDINGS	CONCLUSIONS

1. Wire the circuit shown in the schematic. — —

2. Adjust the SFG to 100 Hz sine waves at 5 volts. — —

3. While measuring the voltage of R_1 (V_{R1}), increase in 100 Hz steps the frequency to the circuit until 1.5 kHz is reached.

   3. Table 1

   $C1 = .1\,\mu F$

Freq. (Hz)	VR1	I tot.	VL1	Z'	Angle
100					
200					
300					
400					
500					
600					
700					
800					
900					
1000					
1100					
1200					
1300					
1400					
1500					

4. Calculate the total current (V_{R1}/R_1) of the circuit at each of the frequencies.

5. Let $Z' = V_{L1}/I$ total. Calculate Z' at each of the frequencies. Note: Z' is the ohmic equivalency of the parallel LC portion of the circuit. — —

6. Using the formula "tangent of the phase angle = $I_C\text{-}I_L/I_{R2}$," calculate the phase angle of Z' at each of the frequencies. — —

7. Calculate the resonant frequency (f_r) of the circuit.

   7. Resonant Frequency = _____ Hz with C1 = .1 µF

   Resonant Frequency = _____ Hz with C1 = .05 µF

   7. Did the resonant frequency change when the capacitor was replaced? _____.
   Did you think it would? _____.
   _____.
   Why? _____
   _____
   _____
   _____.

Experiment 63 Parallel Resonant Circuits ■ 243

PROCEDURES	FINDINGS	CONCLUSIONS

8. Replace the .1 µF capacitor with a .05 uF unit.

8. Table 2

$C1 = .05\,\mu F$

Freq. (Hz)	VR1	I tot.	VL2	Z′	Angle
100					
200					
300					
400					
500					
600					
700					
800					
900					
1000					
1100					
1200					
1300					
1400					

—

9. Repeat steps 3 through 7.

—

9. Was the total current, phase angle, and impedance greatest or least when the circuit was operated at (or near) the resonant frequency?

Current: _____.
Why? _____
_____.

Phase angle: _____.
Why? _____
_____.

Z: _____.
Why? _____
_____.

Was the phase angle - or + when operating below f_r? ___
_____.

Was the phase angle - or + when operating above f_r? ___
_____.

10. Turn off the equipment.

—

—

PROCEDURES	FINDINGS	CONCLUSIONS
11. Compare the two sets of data.	—	11. Did the current increase or decrease as you increased the SFG frequency to the circuit? _____. Was it highest or lowest at resonance? _____. Once the resonant frequency was exceeded, what happened to the current? _____ _____. _____.
—	—	12. Compare the findings of this lab to that of the series resonant circuit, Experiment 61.

Experiment 64

PHASE ANGLE MEASUREMENTS USING THE OSCILLOSCOPE

SCHEMATIC:

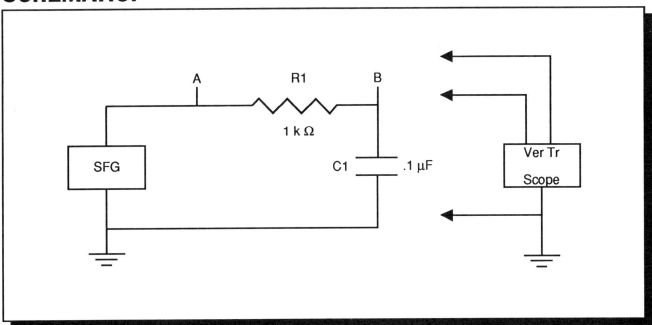

DISCUSSION

The oscilloscope can be used to make a variety of electrical measurements. It can display both AC and DC voltage waveforms from which amplitudes, frequencies, and the waveshapes themselves can be determined. In addition, the scope can be used for the measuring of phase angles. In this experiment we will see how a single trace triggered sweep scope can be used in this manner.

TEXT CORRELATION

Before beginning this experiment, review section 16–7 on phase shifts in *Electricity*.

OBJECTIVE

1. To measure the phase angles of voltages in series RC circuits using a single trace oscilloscope.

EQUIPMENT

- ☐ Standard tool kit
- ☐ Standard parts kit
- ☐ Signal or function generator
- ☐ VOM
- ☐ Single trace triggered sweep oscilloscope with cables (A multiple channel scope may be substituted using one channel).

SAFETY NOTES

 Remember! The screen of the scope is GLASS. Be careful not to strike it as an implosion may result. Remember! Capacitors are capable of holding voltages long after the supply voltage is removed. Be certain they are discharged before handling.

PROCEDURES	FINDINGS	CONCLUSIONS

1. Wire the circuit shown in the schematic. — —

2. Adjust the SFG to 1 kHz and 2 V_{eff} as measured by the VOM. — —

3. The scope should be AC coupled and vertically calibrated. The time base (time per division) should be set to .1 millisecond per centimeter. — —

4. Set the scope for external triggering and the volts per division control to .5 volts per centimeter. — —

5. Connect the vertical input and external trigger input probes of the scope to point A in the schematic and its ground lead to the circuit ground. — —

6. Using the horizontal calibration control, adjust the scope to display a single sine wave. — —

7. By dividing the 360 degrees of 1 cycle by the number of centimeters covered by that cycle, determine the number of degrees represented by each of the major divisions of the CRT screen.

   7. One cycle = _____ centimeters.

   Degrees per centimeter = _____

   7. Does this appear to give the scope an electrical angle measuring capability? _____

8. With the trigger level control, adjust the display to begin the wave at the instant it starts moving in the positive direction from the zero reference line. — —

Experiment 64   Phase Angle Measurements Using the Oscilloscope ■ 247

PROCEDURES	FINDINGS	CONCLUSIONS

9. Record the waveform on Graph 1.

9. Graph 1

—

10. Move the vertical input probe to point B. Leave the other leads connected and scope settings as before.

—

—

11. Record the new waveform also on Graph 1.

—

11. How do the two waveforms compare? _____

Do the two waveforms appear to be "out of phase"?

What phase angle is indicated by the two scope displays?

Explain how you determined this. _____

12. Turn off the equipment.

—

—

13. Calculate the phase angle of this circuit using the formula "tangent of the phase angle = X_C/R."

13. Phase angle from the formula

= _____.

13. Is the phase angle calculated by the formula and that indicated by the scope approximately the same? _____

Does this experiment demonstrate the phase angle measuring capability of the scope?

Experiment 65

DIODE - RESISTIVE CHARACTERISTICS

SCHEMATIC:

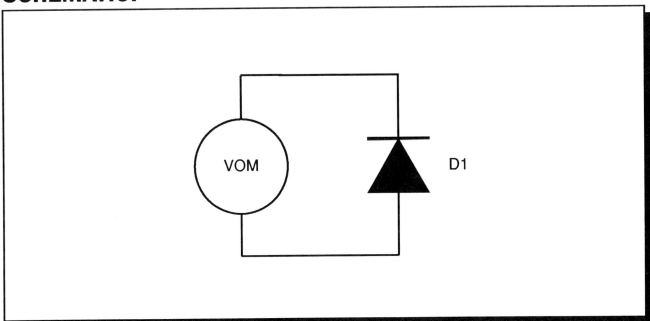

DISCUSSION

A diode is an electrical device that allows current flow in only one direction. This component can be used to provide electrical isolation, rectification, detection, and a number of important circuit operations. In this procedure we will measure the forward biased (front) and reverse biased (back) resistances of a silicon diode and see how they relate to basic diode operation.

TEXT CORRELATION

Before beginning this experiment, review section 21–1 on diodes in *Electricity*.

OBJECTIVES

1. To measure the forward (+ to +) and reverse biased (+ to -) resistances of a silicon diode.

2. To determine the front to back resistive ratio of a solid state diode.

EQUIPMENT

☐ Standard tool kit

☐ Standard parts kit

☐ VOM

249

250 ■ Diode - Resistive Characteristics

PROCEDURES	FINDINGS	CONCLUSIONS
1. Wire the circuit shown in the circuit.	—	—
2. Set the VOM to a low ohms range such as Rx10.	2. Note: If you are using an electronic VOM (vacuum tube, BJT, or FET type) set it to the junction test function.	—
3. Measure and record the resistance of the diode with the positive (red) lead of VOM connected to the anode and the negative (black) connected to the cathode.	3. Diode resistance = _____ ohms.	3. The diode is said to be _____ _____ biased.
4. Set the VOM to a higher ohms range such as Rx100.	—	—
5. Reverse the ohmmeter connections to the diode (positive to the cathode and the negative to the anode).	—	5. The diode is now said to be _____ biased.
6. Again measure and record the resistance readings of the diode.	6. Diode resistance = _____ ohms.	—
7. Using the above measurements calculate the forward to reverse biased resistance ratio.	—	7. This calculation is sometimes referred to as the _____ to _____ or _____ to _____ resistance ratio. For a good silicon diode this ratio should be at least _____ to one. If the diode is said to be "shorted" what effect will this have on the: a. forward resistance? _____ _____ b. reverse resistance? _____ _____ Could these resistive checks be useful in troubleshooting diodes? _____ Explain your answer. _____ _____

Experiment 66

HALF-WAVE RECTIFICATION

SCHEMATIC:

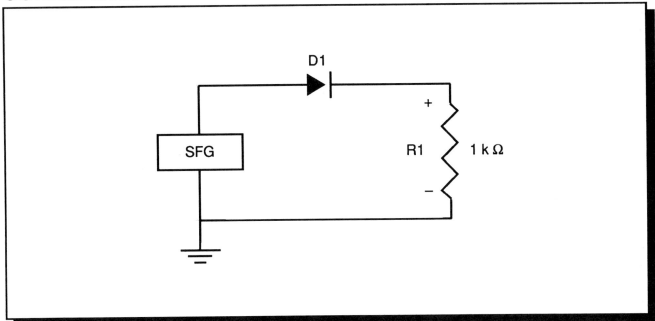

DISCUSSION

A diode is a two-element component that will conduct electricity in only one direction. This device becomes most useful in the process of converting an alternating current (AC) into a direct current (DC), a process known as rectification. In this procedure we will investigate the use of a diode in a half-wave rectifier circuit. The diode will allow only one half of the AC waveform to pass (usually the positive half) while blocking the other (negative) half. Although there will be large fluctuations in the amplitude, the direction of current flow will be prevented from reversing.

TEXT CORRELATION

Before beginning this experiment, review section 21–2 on half-wave rectifiers in *Electricity*.

OBJECTIVES

1. To determine that diodes can convert AC into DC.
2. To examine the half-wave rectifiers' waveforms.

EQUIPMENT

- ☐ Standard tool kit
- ☐ Standard parts kit
- ☐ Signal or function generator (SFG)
- ☐ VOM
- ☐ Triggered sweep oscilloscope

252 ■ Half-Wave Rectification   Experiment 66

SAFETY NOTES:

 Remember! Both AC and DC voltages are present in rectification circuits. Make certain measuring devices are set appropriately.

PROCEDURES	FINDINGS	CONCLUSIONS

1. Wire the circuit shown in the schematic 1. — —

2. Adjust the frequency of the SFG to 100 Hz. and the output voltage to 5 V_{eff}. — —

3. Connect the scope to the "high" output of the SFG. — —

4. Adjust the scope to display approximately three complete cycles. — —

5. Record the image on the scope in Graph 1 and the peak to peak as well as peak voltages.

    5. Graph 1:

    Input Waveform

    [grid]

    Peak to peak voltage = _____

    Peak voltage = _____.

    5. Does the image agree with the waveform set on the SFG? _____.

6. Move the probes of the scope across R_1 and record the image on the scope in Graph 2.

    6. Graph 2:

    R1 Waveform

    [grid]

    6. Is the image that of a rectified AC voltage? _____.
    Explain._____

    How is this image different from that in Graph 1?_____

    Could this voltage now be considered DC? _____.

Experiment 66　　　　　　　　　　　　　　　　　　　　　　　　Half-Wave Rectification ■ 253

PROCEDURES	FINDINGS	CONCLUSIONS
		Why?_____ _____ _____.
7. Set the VOM to DC volts. Measure and record the voltage across R_1.	7. Measured DC voltage of R_1 = _____.	—
8. Turn off the equipment.	—	—
9. Calculate the VDC that should have resulted by multiplying the measured VAC peak voltage by .318.	9. Calculated DC voltage of R_1 = _____.	9. Does the calculated DC voltage approximately agree with that measured in step 7?____ _____.
		If not, explain._____ _____ _____.
		It can be concluded that a _____ passes a current in only one direction, thereby giving it the capability of converting an _____ into a _____.

Experiment 67

FULL-WAVE BRIDGE RECTIFIER

SCHEMATIC:

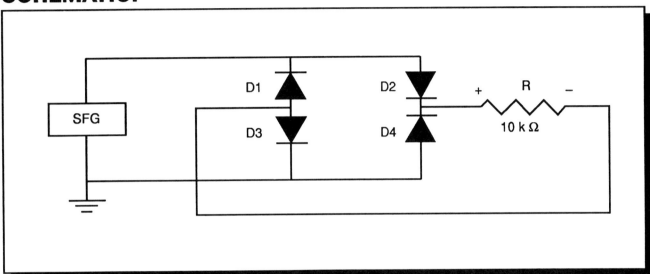

DISCUSSION

A diode is a two-element component that will conduct electricity in only one direction. This device becomes most useful in the process of converting an alternating current (AC) into a direct current (DC) a process known as rectification. In this procedure we will investigate the use of a diode in a full-wave bridge rectifier circuit. As with the half-wave rectifier of experiment 66, this circuit allows only voltages of one polarity to appear across the load. Although there are large fluctuations in the amplitude, the polarity is prevented from reversing. Since the bridge rectifier costs less to produce and provides higher output values than a full wave rectifier (using two diodes and a center-tapped transformer as power source) it is the most commonly used.

TEXT CORRELATION

Before beginning this experiment, review section 21–2 on half-wave rectifiers in *Electricity*.

OBJECTIVES

1. To determine that diodes can convert AC voltages into DC.

2. To examine the full-wave rectifiers' output waveform and voltage.

EQUIPMENT

- ☐ Standard tool kit
- ☐ Standard parts kit
- ☐ Signal or function generator (SFG)
- ☐ Triggered sweep oscilloscope
- ☐ VOM

SAFETY NOTES

 Remember! Both AC and DC voltages are present in rectification circuits. Make certain measuring devices are set appropriately.

PROCEDURES	**FINDINGS**	**CONCLUSIONS**
1. Wire the circuit shown in the schematic 1.	—	—
2. Adjust the frequency of the SFG to 100 Hz. and the output voltage to 5 V_{eff}.	—	—
3. Connect the scope to the "high" output of the SFG.	—	—
4. Adjust the scope to display approximately three complete cycles.	—	—
5. Record the image on the scope in Graph 1 and the peak to peak as well as peak voltages.	5. Graph 1 **Input Waveform** Peak to peak voltage = _____. Peak voltage = _____.	5. Does the image agree with the waveform setting of the SFG? _____.
6. Move the probes of the scope across R_1 and record the image on the scope in Graph 2.	6. Graph 2 **R1 Waveform**	6. Is the image that of a rectified AC voltage? _____. Explain. _____. How is this image different from that in Graph 1? _____ Could this voltage now be considered DC? _____ Why? _____.

Experiment 67 Full-Wave Bridge Rectifier ■ 257

PROCEDURES	FINDINGS	CONCLUSIONS
		How does this waveform compare to that displayed in step 6 of experiment 66? ____
7. Set the VOM to DC volts. Measure and record the voltage across R_1.	7. Measured DC voltage of R_1 = _____.	7. How does this voltage compare to that measured in step 7 of experiment 66? _____
8. Turn off the equipment.	—	—
9. Calculate the VDC that should have resulted by multiplying the measured VAC peak voltage by .636.	9. Calculated DC voltage of R_1 = _____.	9. Does the calculated DC voltage approximately agree with that measured in step 7? ____ If not, explain. _____
10. Examine the circuit.	—	10. When the ungrounded output terminal of the SFG is positive, the current first passes thru diode D_____, then the load resistor (R_1), then through diode D_____ and finally back to the other side of the source. When the ungrounded output terminal of the SFG is negative, the current first passes through diode D_____, then R_1, then diode D_____, then back to the other side of the source. Does the current pass through R_1 in the same direction regardless of the polarity of the source? _____. It can be concluded that this circuit allows current flow in only one direction, thus producing a _____ output (load) voltage.

Experiment 68

HALF-WAVE RECTIFIER WITH CAPACITOR FILTER

SCHEMATIC:

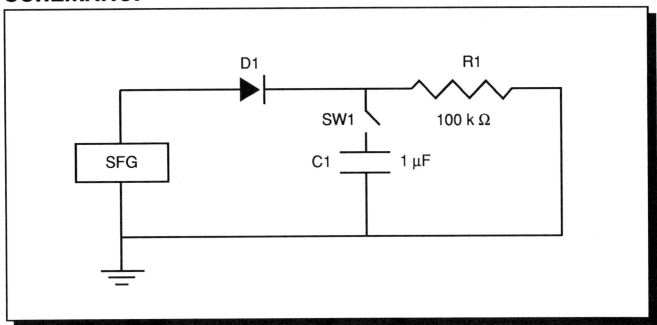

DISCUSSION

The diode is a device that can be used to convert AC to DC. The DC voltage that it creates is of a pulsating type. For many applications this is not acceptable. A "filtering" component such as a capacitor may be added to the output of the circuit to reduce the amplitude of these pulses, thus providing a more constant level of DC voltage.

TEXT CORRELATION

Before beginning this experiment, review section 21–2 on half-wave rectifiers in *Electricity*.

OBJECTIVE

1. To examine the use of the capacitor as a DC filter.

EQUIPMENT

- ☐ Standard tool box
- ☐ Standard parts kit
- ☐ Signal or function generator (SFG)
- ☐ VOM
- ☐ Triggered sweep oscilloscope with cables

260 ■ Half-Wave Rectifier Filter with Capacitor Filter   Experiment 68

SAFETY NOTES

 Remember! The capacitor is capable of holding a voltage long after being removed from the supply. Be certain the capacitor is fully discharged before handling.

PROCEDURES	**FINDINGS**	**CONCLUSIONS**

1. Wire the circuit shown in the schematic. — —

2. Obtain a sharp horizontal line on the vertical center of the screen. Then set the scope for AC coupling. — —

3. Check to be certain that the scope is in a vertical input (voltage) calibrated mode. Then connect it to the "high" output of the SFG. — —

4. Adjust the SFG to 100 Hz sinewaves at a 5 volts effective. — —

5. Adjust the scope to display approximately three complete cycles. —

6. With the scope connected to the output of the SFG observe and draw the pattern appearing on the scope on Graph 1.

   6. Graph 1:

   Input Waveform

   6. Does the waveform on the scope agree with the waveform setting indicated on the function generator? _____.

7. Reconnect the scope and the VOM across R_1 with switch 1 (SW1) open (off). — —

8. Calibrate and then record the waveform of the scope on Graph 2.

   8. Graph 2:

   Output Waveform Without Filter

   8. How do the waveforms in Graphs 1 and 2 differ? _____
   _____
   _____
   _____.

Experiment 68   Half-Wave Rectifier Filter with Capacitor Filter ■ 261

PROCEDURES	FINDINGS	CONCLUSIONS

9. Measure and record in Table 1 the DC (VOM) and the AC (scope) voltages across R1. Remember, the AC input has been rectified; therefore, adjust the VOM for DC measurements.

9. Table 1

SW1	DC Volts (VOM)	AC Volts (Scope)	% Ripple
Open			
Closed			

—

10. Calibrate Graph 3.

10. Graph 3:

Output Waveform With Filter

—

11. With the scope and VOM still connected to R_1, close SW1 and record the new waveform on Graph 3 and the DC voltage in Table 1.

—

11. When the capacitor was added to the output of the circuit (SW1 closed), the DC voltage _____ and the AC (ripple) voltage _____

How do the waveforms, with (Graph 3) and without (Graph 2) the filter capacitor, compare? _____
_____.

12. Return the SFG voltage to 0 and turn off the equipment.

—

—

13. Calculate and record in Table 1 the % of ripple of the circuit with and without the filter connected.

—

13. When the capacitor was switched into the circuit, the % of ripple _____
_____.

It can be concluded that a _____ aids the rectifier in creating a more constant and battery-like DC voltage.

Experiment 69

BIPOLAR TRANSISTOR - RESISTIVE CHARACTERISTICS

SCHEMATIC:

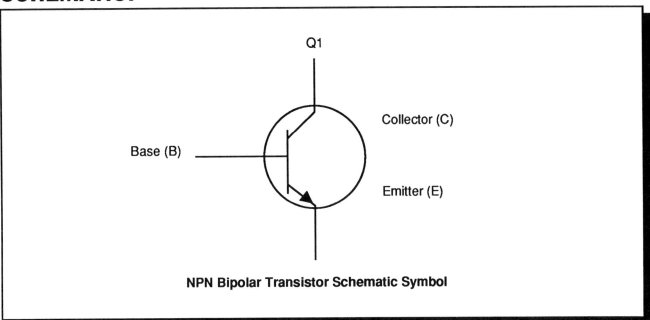

NPN Bipolar Transistor Schematic Symbol

DISCUSSION

The bipolar transistor, sometimes referred to as the bijunction transistor (BJT), is among the first of the solid state active devices. It was the component that launched the era of modern micro-electronics. The significance of this component is far too great to be done justice in this small space. Suffice it to say, the state of present-day electronics would be inconceivable without the invention of this component.

TEXT CORRELATION

Before beginning this experiment, review section 21–3 on current operated transistors in *Electricity*.

OBJECTIVES

1. To measure the resistance of a silicon bipolar transistor.

EQUIPMENT

- ☐ Standard tool kit
- ☐ Standard parts kit
- ☐ VOM
- ☐ Transistor data sheet

SAFETY NOTES

 Remember! BJTs, like all semiconductors, should be handled with care. They are vulnerable to damage from static charges as well as lead breakage.

PROCEDURES	**FINDINGS**	**CONCLUSIONS**
1. Using the transistor data sheet, identify the leads corresponding to the emitter (E), base (B), and collector (C) of the transistor (Q_1).	—	1. Since an NPN transistor is being used, the _____ and _____ are made of an N-type semiconductor material and the _____ is made of a P-type.
2. Set the VOM to the Rx10 scale. If you are using an electronic VOM, such as an FET or digital type, use the junction test setting.	—	2. Note: The internal source of the VOM will provide DC biasing for the PN junctions.
3. Connect the positive lead of the meter to the base and the negative lead to the emitter of the transistor.	—	—
4. Record the reading in Table 1.	4. Table 1	—

Ohmmeter Connections		Ohms	Biasing
+ Base	– Emitter		
– Base	+ Emitter		
+ Base	– Collector		
– Base	+ Collector		
+ Collector	– Emitter		
– Collector	+ Emitter		

5. Reverse the connections of the VOM leads to the transistor and record the new readings in Table 1.	—	—
6. Reconnect the VOM to the base and collector of the transistor and repeat steps 4 and 5.	—	—
7. Reconnect the VOM to the collector and the emitter of the transistor and again repeat steps 4 and 5.	—	—
8. Turn off the VOM.	—	—

PROCEDURES	FINDINGS	CONCLUSIONS
9. Examine Table 1.	—	9. A PN junction is said to be forward biased if a more positive voltage is on the _____ material and a more negative voltage is on the _____ material.
10. In the biasing column of Table 1, indicate if the PN junctions were being forward or reverse biased by each of the VOM to transistor connections.	—	10. It can be concluded that when a PN junction is forward biased it has a _____ resistance, and when it is reversed biased it has a _____ resistance.

Experiment 70

BIPOLAR TRANSISTOR — DC OPERATION

SCHEMATIC:

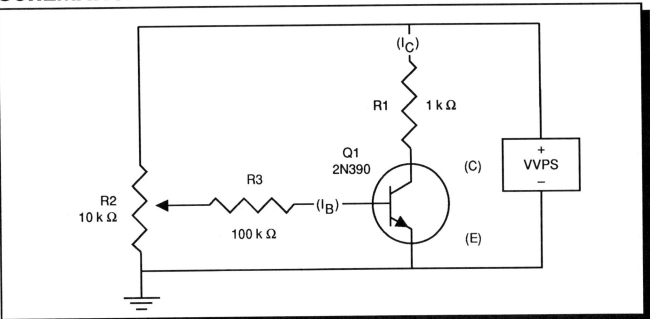

DISCUSSION

The bipolar transistor, sometimes referred to as the bijunction transistor (BJT), is among the most useful of solid state devices. It is classified as an "active device" as it is capable of controlling an electric current electrically. That is to say, one current is able to control another current within the confines of this component.

TEXT CORRELATION

Before beginning this experiment, review section 21–3 on current operated transistors in *Electricity*.

OBJECTIVES

1. To measure the the relationship between base current (I_B) and collector current (I_C) in a bipolar transistor.

EQUIPMENT

- ☐ Standard tool kit
- ☐ Standard parts kit
- ☐ Two VOM's
- ☐ Variable voltage DC power supply (VVPS)

SAFETY NOTES

 Remember! Transistors are very sensitive to their voltage or current ratings being exceeded. If this occurs, permanent damage to the device will result.

PROCEDURES	FINDINGS	CONCLUSIONS

1. Wire the circuit shown in in the schematic. — —

2. Set the VVPS to 10 volts DC. — —

3. Set the I_B VOM to the 1 mA DC range and the I_C VOM to the 10 mA range. — —

4. Rotate the potentiometer (R_2) fully counterclockwise. — —

5. Record the base and collector current readings in Table 1.

5. Table 1

Base Current (I_B)	Collector Current (I_C)
.0 mA	
.1 mA	
.2 mA	
.3 mA	
.4 mA	
.5 mA	
.6 mA	
.7 mA	
.8 mA	
.9 mA	
1.0 mA	

5. This current should have been very low. Why? _____

6. Begin slowly rotating R_2 in the clockwise direction until I_B is equal to .1 mA. —

6. As the potentiometer is being rotated in the CCW direction, the voltage from it is _____

7. Record the new value I_C in Table 1. — —

8. Continue to increase the base current in .1 mA increments until 1 mA is reached. —

8. As the base current increased, the collector current _____

PROCEDURES	FINDINGS	CONCLUSIONS
		Why? _____ _____ _____
9. Turn off the equipment.	—	—
10. Calibrate Graph 1 and plot the data on Table 1.	10. Graph 1 I_C(mA) vs I_B(mA) with markings at .2 mA, .6 mA, 1.0 mA	10. The graph of output current, voltage, or power versus input current, voltage, or power is common for all types of active components. It is often referred to as a _____ _____ curve.

Experiment 71

BIPOLAR TRANSISTOR SIGNAL AMPLIFIER

SCHEMATIC:

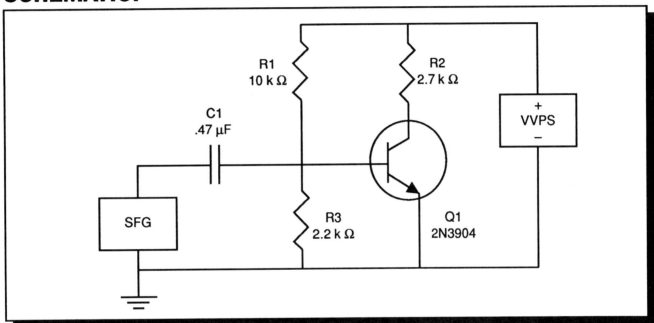

DISCUSSION

The bipolar transistor (BJT) is one of the most widely used solid state active devices. It is found in both "discrete" and "integrated" circuitry. The BJT is a current controlled device that operates by using combinations of forward and reverse biased PN junctions. It can function as either a switching (digital) or a regulating (amplifying) component.

TEXT CORRELATION

Before beginning this experiment, review section 21–3 on current operated transistors in *Electricity*.

OBJECTIVES

1. To examine the use of the BJT as an AC signal amplifier.

EQUIPMENT

- ☐ Standard tool kit
- ☐ Standard parts kit
- ☐ Signal or function generator (SFG)
- ☐ Triggered sweep oscilloscope
- ☐ Variable voltage DC power supply (VVPS)

271

272 ■ Bipolar Transistor Signal Amplifier

Experiment 71

SAFETY NOTES

Remember! Transistors are very sensitive to their voltage or current ratings being exceeded. So doing will cause permanent damage.

PROCEDURES	FINDINGS	CONCLUSIONS

1. Wire the circuit shown in the schematic. — 1. What is the purpose of C_1 in this circuit? _____

2. Set the VVPS to 10 volts DC. — —

3. Set the oscilloscope for AC coupling and connect it to the base of Q_1. — —

4. Set the SFG for a sine wave output at a frequency of 1 kHz and a voltage of .2 volts peak-to-peak. — —

5. Reconnect the scope to the collector of the BJT and record the peak-to-peak voltage in Table 1.

5. Table 1:

AC Base Voltage	AC Collector Voltage	AC Voltage Gain
.2 Volts P-P		
.4 Volts P-P		
.6 Volts P-P		
.8 Volts P-P		
1.0 Volts P-P		

5. Is the AC voltage on the collector larger than that on the gate? _____.

6. While repeating steps 3 through 5 for each of the input signal voltages, increase the output of the SFG in .2 volt increments until 1 volt peak-to-peak is reached.

—

6. Did the AC voltage at the collector increase as the AC voltage at the base was increased? _____.

Why? _____

7. Turn off the equipment. — —

8. Calculate and record in Table 1 the AC voltage gain of the circuit for each of the values of input (base) voltage.

8. Voltage gain (A_V) is equal to the _____ voltage divided by the _____ voltage.

9. Calibrate Graph 1 and then plot on it the collector and base voltages recorded in Table 1.

9. Graph 1

V Collector (p-p)

.4V   .8V   V Base (Vp-p)

9. This type of graph, output voltage versus input voltage, is sometimes referred to as a _____ curve or function.

PROCEDURES	FINDINGS	CONCLUSIONS
		When the AC voltage at the base is at .7 V_{p-p}, the AC voltage at the collector is _____.
		Was the AC voltage gain (A_v) of the circuit relatively constant over the range of input voltages? _____.
		It can be concluded that the BJT may indeed be used as a _____ amplifier.

Experiment 72

JFET-RESISTIVE CHARACTERISTICS

SCHEMATIC:

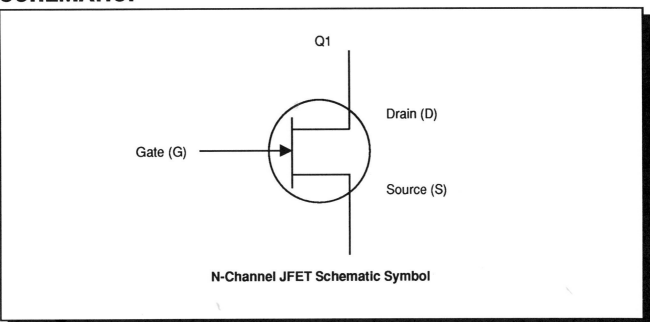

N-Channel JFET Schematic Symbol

DISCUSSION

Another within the family of solid state active devices is the junction field effect transistor (JFET). The JFET can be differentiated from the bipolar transistor (BJT) by the fact that it is a voltage operated device as opposed to a current operated device.

TEXT CORRELATION

Before beginning this experiment, review section 21–4 on voltage operated transistors in *Electricity*.

OBJECTIVE

1. To measure the DC resistance of an N-channel JFET.

EQUIPMENT

- ☐ Standard tool kit
- ☐ Standard parts kit
- ☐ VOM
- ☐ Junction field effect transistor data sheet

SAFETY NOTES

 Remember! JFETs, like all semiconductors, should be handled with care. They are very vulnerable to damage from static charges as well as lead breakage.

PROCEDURES | FINDINGS | CONCLUSIONS

1. Using the transistor data sheet identify the leads corresponding to the gate (G), source (S), and drain (D) of the transistor (Q_1).

   —

   1. Since an N-channel JFET is being used, the _____ and _____ are made of an N-type semiconductor material and the _____ is made of a P-type.

2. Set the VOM to the Rx10 scale. If you are using an electronic VOM, such as an FET or digital type, use the junction test setting.

   —

   —

3. Connect the positive lead of the meter to the gate and the negative lead to the source of the transistor.

   —

   3. Note: The PN junctions are being biased by the internal source of the ohmmeter (VOM).

4. Record the reading in Table 1.

   4. Table 1

Ohmmeter Connections		Ohms	Biasing
+ Gate	– Source		
– Gate	+ Source		
+ Gate	– Drain		
– Gate	+ Drain		
+ Drain	– Source		
– Drain	+ Source		

   4. Was the resistance from the drain to the source low regardless of the polarity of the VOM? _____.

   Why? _____ _____.

5. Reverse the connections of the VOM leads to the transistor and record the new readings in Table 1.

   —

   —

6. Reconnect the VOM to the gate and drain of Q_1 and repeat steps 4 and 5.

   —

   —

7. Reconnect the VOM to the drain and the source of the transistor and again repeat steps 4 and 5.

   —

   —

PROCEDURES	FINDINGS	CONCLUSIONS
8. Turn off the VOM.	—	—
9. By examining the ohms column of Table 1, indicate if the regions were being forward or reverse biased by each of the VOM to transistor connections.	—	9. A PN junction is said to be forward biased if a more positive voltage is on the _____ material and a more negative voltage is on the _____ material. It can be concluded that when a region of a JFET is forward biased it has a _____ resistance, and when reversed biased it has a _____ resistance.

Experiment 73

JFET—DC OPERATION

SCHEMATIC:

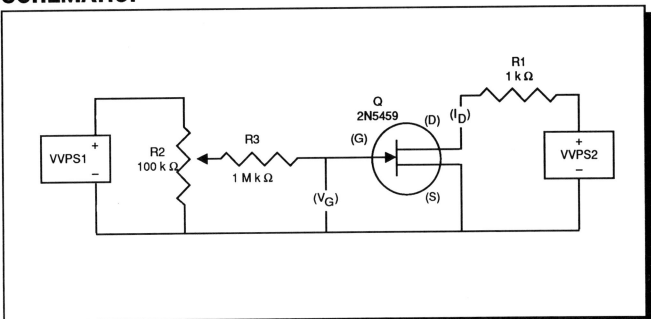

DISCUSSION

The JFET is a voltage controlled active device. This means that voltages placed on the input (gate) are able to control the amount of current passing through the channel. This control is so great that the gate voltage is capable of stopping the source-drain current completely. This condition is sometimes referred to as "pinchoff".

TEXT CORRELATION

Before beginning this experiment, review section 21–4 on voltage operated transistors in *Electricity*.

OBJECTIVE

1. To examine the relationship of gate voltage to channel current in a JFET.

EQUIPMENT

- ☐ Standard tool kit
- ☐ Standard parts kit
- ☐ VOM
- ☐ Two variable voltage DC power supplies (VVPS)

280 ■ JFET – DC Operation   Experiment 73

SAFETY NOTES

 Remember! Solid state devices are very sensitive to their voltage or current ratings being exceeded. If this occurs, permanent damage to the device will result.

PROCEDURES	FINDINGS	CONCLUSIONS

1. Wire the circuit shown in the schematic.

   —

   1. Is the gate of the JFET being forward or reverse biased? _____

   Does that help it achieve its characteristic high input impedence? _____.

2. Set the drain current (I_D) VOM to the 10 mA DC range.

   —

   —

3. Set the gate voltage (V_G) VOM to the 10 volts DC range.

   —

   —

4. Rotate R_2 to the fully clockwise position.

   —

   —

5. Record in Table 1 the readings on the VG and the drain current (I_D) VOMs.

   5. Table 1

Gate Voltage (V_G)	Drain Current (I_D)
Volts	
.4 Volts	
.8 Volts	
1.2 Volts	
1.6 Volts	
2.0 Volts	
2.4 Volts	
2.8 Volts	
3.2 Volts	
3.6 Volts	
4.0 Volts	
4.4 Volts	
4.8 Volts	
5.2 Volts	
5.6 Volts	
6.0 Volts	

   5. Was this voltage approximately zero? _____.
   Should it have been? _____.
   Why? _____
   _____
   _____.

   Was the drain current relatively high at this gate voltage? _____
   _____.

   Why? _____
   _____
   _____.

PROCEDURES	FINDINGS	CONCLUSIONS
6. Slowly rotate R_2 in the clockwise direction until V_G measures .4 volts. Record the drain current (I_D) in Table 1.	—	—
7. Increase the V_G in .4 volt increments until 6 volts is reached. Record the I_D in Table 1 for each of the V_G settings.	—	7. Did the drain current decrease as the negative gate voltage was increased? _____ _____. Was that to be expected? ____. Why? _____ _____ _____. As the gate voltage was increased, did the drain current stop flowing at some point? _____. If so, list that voltage here _____ volts. This voltage is referred to as the _____ voltage.
8. Turn off the equipment.	—	—

Experiment 74

JFET-VOLTAGE AMPLIFIER

SCHEMATIC:

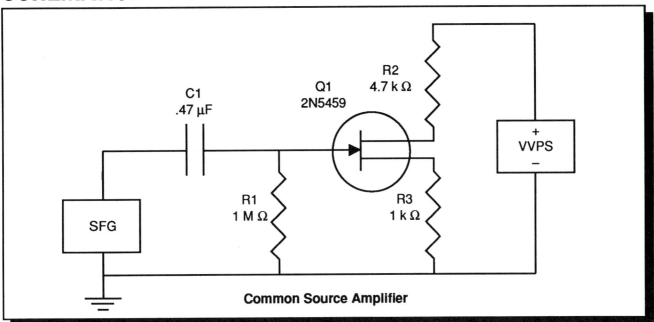

Common Source Amplifier

DISCUSSION

The junction field effect transistor (JFET) is a voltage operated active device as opposed to a current operated component. The input to the circuit above is the gate. As this is a reverse biased material, only a small amount of current will be drawn from the signal source. Due to this, the device produces very little signal "loading".

TEXT CORRELATION

Before beginning this experiment, review section 21–4 on voltage operated transistors in *Electricity*.

OBJECTIVES

1. To examine the use of the JFET as a signal voltage amplifier.

EQUIPMENT

- ☐ Standard tool k
- ☐ Standard parts kit
- ☐ Variable voltage DC power supplies (VVPS)
- ☐ Signal or sweep generator (SFG)
- ☐ Triggered sweep oscilloscope

284 ■ JFET – Voltage Amplifier

Experiment 74

SAFETY NOTES

  Remember! Transistors are very sensitive to their voltage or current ratings being exceeded. To do so will cause them to become permanently damaged.

PROCEDURES	**FINDINGS**	**CONCLUSIONS**

1. Wire the circuit shown in the schematic.

1. What is the purpose of C_1 in this circuit?

—

2. Set the oscilloscope for AC coupling and connect it to the drain of the gate of the JFET.

—

—

3. Set the SFG for a sinewave output at a frequency of 1 kHz and a voltage of .2 volts peak-to-peak.

—

—

4. Set the VVPS to 10 volts DC.

—

—

5. Connect the oscilloscope to the drain of the JFET and record the peak-to-peak drain voltage in Table 1.

5. Table 1:

AC Gate Voltage	AC Drain Voltage	AC Voltage Gain
.2 V_{P-P}		
.4 V_{P-P}		
.6 V_{P-P}		
.8 V_{P-P}		
1.0 V_{P-P}		

5. Is the AC voltage on the drain larger than that on the gate?

_____

6. While repeating step 5 for each of the input signal (AC) voltages, increase the output of the SFG in .2 volt peak-to-peak increments until 1 volt is reached.

—

6. Did the AC voltage at the drain increase as the AC voltage at gate was increased?

_____

Why? _____

_____

_____.

7. Turn off the equipment.

—

—

8. Calculate and record in Table 1 the AC voltage gain of the JFET for each of the values of input (gate) voltage.

—

8. Voltage gain (A_v) is equal to the _____ voltage divided by the _____ voltage.

PROCEDURES	FINDINGS	CONCLUSIONS
9. Calibrate Graph 1, then plot the V_D and V_G data of Table 1.	9. Graph 1 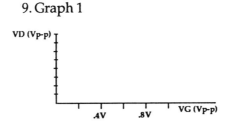	9. This type of graph (output voltage versus input voltage) is sometimes referred to as a _____ curve or function. When the AC voltage at the gate is .7 V_{p-p}, the AC voltage at the drain is _____. Was the AC voltage gain (A_v) of the circuit relatively constant over the range of input voltages? _____. It can be concluded that the JFET may indeed be used as a _____ amplifier.